Gabriel Caymmi Vilela Ferreira

Rural Settlements in Mato Grosso's Araguaia Valley

Gabriel Caymmi Vilela Ferreira

Rural Settlements in Mato Grosso's Araguaia Valley

Adaptation and Permanence

ScienciaScripts

This book is a translation from the original published under ISBN 978-613-9-62376-1.

Publisher:
Sciencia Scripts
is a trademark of
Dodo Books Indian Ocean Ltd. and OmniScriptum S.R.L publishing group

120 High Road, East Finchley, London, N2 9ED, United Kingdom
Str. Armeneasca 28/1, office 1, Chisinau MD-2012, Republic of Moldova, Europe
Printed at: see last page
ISBN: 978-620-7-72439-0

SUMMARY

I dedicate this work to every rural man who has been wronged by the selective and classifying politics that sees only large rural enterprises as the way forward for agriculture.

To agroecological farmers and rural social movements, because this is the future I want for our agriculture.

ACKNOWLEDGMENTS

Thanking the people who were part of this achievement is both a joy for me and an injustice to those who will not be remembered in these lines, but who were fundamental to making this work happen. Therefore, my first thanks go to those who will not be remembered, either here in these lines or in everyday life. My thanks and my wish for change go to them.

I would like to thank my father, Kiko França, for his support in carrying out this research, not only the financial support (which was essential, the "check" for the whole month), but also the support in liaising with the settlers, making the structure available at INCRA and his fatherly support. I would like to thank my mother Maria José (Coroa) who is my mainstay, who always helps me even when she thinks she's not helping, who always brings me peace and affection and is the source of my constant laughter. Thank you to my brother Pedro Rafael, for being the example I always try to follow - he is both my friend and my idol.

I would like to thank Professor Ambrósio. In the more than two years we have been together, I have never found the door to his office closed, and it is with this symbolic gesture that he transforms the act of mentoring into something more important. To the Settlement research group, for the discussions, the partnership, the brothers and sisters that we have forged in our daily toil as researchers. Super Rose and Poli for their sincere help. To the Department of Rural Economy, the first place at UFV where I felt I belonged, included.

I would like to thank Romildo for his "co-supervision". He has long since ceased to be just the postgraduate secretary and is now a tutor/friend to the master's and doctoral students who pass through his office. To Carminha and Margarida for their support.

I would like to thank the Rural Extension postgraduate class of 2013: you were fundamental in the difficult times and especially in the times of joy. "We burned a lot of meat and drank a lot of beer. I would especially like to thank Amàbile, my partner at all times; Fernandinha, my mentoring sister; Manu, the gringo I taught to live in Viçosa; Aline, who is from the 2014 master's program, but belongs to us from 2013; and Tiziu, who helped me get into the graduate program.

I would like to thank UAVA/INCRA in Barra do Garças for welcoming me and helping me with the research, especially the agronomist Joao Leao, who showed me the institution's qualities as well as its flaws. Heartfelt thanks to all fifty families who welcomed me into their homes and treated me like a close friend. Thank you to the settlers of Santa Emilia, Volta Grande, Ilha do Coco and Martins I, for the opportunity to understand in practice the difficulties of Brazilian agrarian policy.

I would like to thank LaboraTe and the University of Santiago de Compostela for welcoming me and teaching me so much. To the Galician people and Lugo, for welcoming me like a brother, to

Carlos, Diego, Yolanda, Manuel, Alberto, Krystel, Jesus, Hektor, Asia, Guido and the whole Palo Selfi team, for making my life in Spain happier and more joyful. Special thanks to Nathalia Thais, who helped me before, during my stay in Spain and even now, while I'm writing these thanks, because I'm using her dissertation as an example. Part of this achievement is yours.

Thank you to the entire Quantum Gis community on the internet: I know that this tool is already helping in research and in the real transformation of many people's lives. Thanks to Jorge Santos for his patience and help with the geoprocessing tools and QGIS.

I would like to thank Cissa for her companionship and support in tense moments, and for her affection and calmness when I needed it most.

Thanks to CAPES for the research grant.

I would like to thank my friends and companions in the daily struggle for political activism.

And thank UTOPIA, that inverse logic, of seeking what is out of reach, but which is fundamental, because it keeps us moving forward!

BIOGRAPHY

Gabriel Caymmi Vilela Ferreira, son of Maria José Vilela Ferreira and Joaquim Francisco Ferreira, was born on September 4, 1989, in Barra do Garças - MT.

At the age of seventeen he entered the Agronomic Engineering course at the Federal University of Viçosa, graduating in May 2013.

During his undergraduate studies, he took part in the Academic Center of the Agronomy course, where he gained a deeper understanding of politics and its importance in academic training. In 2011, he became general coordinator of the Central Student Directory - DCE, during the V!raçao administration. This was a defining experience in his formation as an agronomist and as a political activist, and was decisive in his decision to do a master's degree in the area of Rural Extension, with a focus on agrarian reform policy.

In 2013, he joined the Postgraduate Program in Rural Extension, beginning his career as a researcher. He had the opportunity to do an exchange program at the University of Santiago de Compostela - USC, where he did an internship at the Territory Laboratory - LaboraTe.

SUMMARY

FERREIRA, Gabriel Caymmi Vilela, M. Sc., Federal University of Viçosa, June 2015. **Rural Settlements in the Araguaia Valley of Mato Grosso: Adaptation and Permanence.** Advisor: José Ambrósio Ferreira Neto.

This paper analyzes the characteristics that determine the profile of beneficiaries of the Brazilian policy who remain in rural settlements, and who persist in the projects, despite serious problems such as poor infrastructure, lack of support from the state and lack of financial credit. In the midst of a significant proportion of beneficiaries who leave these projects, the reasons for their permanence need to be better clarified. The work was carried out in the Santa Emilia, Volta Grande, Ilha do Coco and Martins I settlements, located in the Araguaia Valley region of Mato Grosso. Structured questionnaires were applied in these settlements during August, September and October 2014. The characteristics of permanence are associated with the pluriactivity of the beneficiaries. They are also not limited to activities focused on agriculture within the settlement, but on various activities to supplement their income. In this way, access to a fixed and continuous income allows these workers to remain in the settlement. In addition, satellite images were used to compare pasture degradation in the initial and current periods of each settlement. As a result, it was possible to see that the cattle management practiced by the settlers had little impact on the environment, promoting recovery from pasture degradation and ratifying more environmentally sustainable practices. It was therefore clear that those who remain in the rural settlements are associated with income from outside the plot and with less intensive and environmentally exploitative farming activities.

INTRODUCTION

The approach taken in this paper to rural government land reform settlements was from the perspective of the permanence of rural workers in these areas. The region analyzed was the Araguaia Valley in the state of Mato Grosso, one of the largest states in the federation, with economic development centered on agribusiness. The literature on the subject exposes the limitations of the settlements, with precarious infrastructure and no support from the public authorities, which leads to considerable evasion. However, there are those who, despite all the difficulties experienced and all the setbacks that this type of policy entails, remain on their plots, producing and surviving in the countryside.

The settlements evaluated were PA Santa Emilia, in the municipality of Barra do Garças; PA Volta Grande, in the municipality of Araguaiana; PA Martins I, in the municipality of Agua Boa; and PA Ilha do Coco, in the municipality of Nova Xavantina, all under the jurisdiction of the Araguaia Valley Advanced Unit (UAVA) of the National Institute for Colonization and Agrarian Reform (INCRA) in Mato Grosso (SR-13).

The guiding question of the research was the interest in understanding that amid precarious situations, such as those experienced by the beneficiaries of the settlement policy, with poor infrastructural conditions and the absence of public power, there are people who flee the projects and others who remain; therefore, who are these people and why do they remain in the rural settlement projects? In view of these questions, the general objective was to identify the profile of rural workers in the settlement projects of the Araguaia Valley in Mato Grosso.

This work consists of four chapters. In the first, the aim was to build a debate about Brazil's land structure over the years and how this process has led to a very large concentration of land in the country. As a result, a portion of the rural population was excluded, creating the need for a reorganization of the Brazilian land structure. One of these alternatives was the National Land Reform Policy - PNRA. However, this chapter also discusses points that problematize this policy and the way in which it has become sectorized, focusing only on the creation of rural settlements which, for the most part, do not even have minimal infrastructure conditions.

The second chapter systematically presents the land question in Mato Grosso, which over the decades has formed a highly concentrated structure. It will also look at the official colonization policy, which promoted the occupation of the Brazilian Midwest in the 1970s and 1980s, with the aim of occupying the existing demographic voids, as well as securing possession of the entire national territory. This chapter will also deal with rural settlements in the state, analyzing their conditions, the difficulties of applying the policy and the evasion of projects.

The third chapter presents in detail the results of the field research carried out using questionnaires applied to residents of the settlement projects mentioned. In this section, the social characteristics of the beneficiaries are analyzed, such as their origin, how they entered the settlement, their current residence and the composition of their family group; in addition, the economic characteristics are also presented, such as the source and origin of income from the plots, external income and main occupations. Productive characteristics will also be covered, highlighting each settler's form of production and the products sold, including those intended only for consumption and identifying the use of labor and the places where this production is sold. The characteristics of the infrastructure - such as schools, health centers, housing and the plot - will also be analyzed. Finally, the issue of dropouts will be addressed, as well as the profile of permanent residents in the settlement projects.

The fourth section presents land use and occupation in the settlements surveyed. Using satellite images collected during the initial period when each project was created, as well as during the current period, the process of pasture degradation was analyzed, classifying them in terms of their degradation intensity. The discussion presented in this chapter was not foreseen in the initial research proposal, but it is an original and technically relevant approach to socio-economic dynamics and its interface with environmental issues in rural settlements.

Methodologically, the work was divided into two phases. The first consisted of applying structured questionnaires to sampled residents of the settlements surveyed, with the aim of collecting documentary material of a higher quality than the data provided by INCRA through SIPRA - Sistema de Informaçao de Projetos de Reforma Agrària[5] . This questionnaire was used to outline the beneficiary's profile, identifying their origin, place of birth, previous experience, jobs performed and their main sources of income, so that this set of characteristics would allow us to analyze and compare the similarities and differences with former beneficiaries of these land reform projects[6] . The sampling process for applying the questionnaires in the four settlements was calculated on a probabilistic basis, using the criteria recommended for finite populations. Thus, taking into account the homogeneity of the population living in the settlements analyzed, we worked with a Degree of Confidence of around 90%, which implied an α of 0.10 which, in turn, implies a Critical Value of $Z_{\alpha 2}$ of 1.645.

In this sense, the equation[7] that made it possible to calculate the sample (n) for the application of the questionnaires in PA Santa Emilia, PA Volta Grande, PA Ilha do Coco and PA Martins I, based on the estimate of the population proportion, was based on the following variables:

---

5 SIPRA is INCRA's internal database, which holds information on all beneficiaries and ex-beneficiaries of this policy.
6 This comparison will be made by collecting data from former beneficiaries through SIPRA - INCRA (secondary data).
7 LEVIN, Jack. Estatistica Aplicada a Ciências Humanas. 2nd Ed. Sao Paulo: Editora Harbra Ltda, 1987.

n = Number of individuals in the sample

N = Population size

$\mathbf{Z}_{\alpha}$ j2 = Critical value corresponding to the desired degree of confidence.

p = Population proportion of individuals belonging to the category we are interested in studying.

q = Population proportion of individuals who do NOT belong to the category we are interested in studying (q = 1 - p).

E = Margin of error or maximum estimation error. Identifies the maximum difference between the sample proportion and the true population proportion (p).

$$n = \frac{N \cdot \hat{p} \cdot \hat{q} \cdot \left(Z_{\alpha/2}\right)^2}{\hat{p} \cdot \hat{q} \cdot \left(Z_{\alpha/2}\right)^2 + (N-1) \cdot E^2}$$

After this first stage for the sample calculation, taking into account the small size of the original population, this initial sample was recalculated considering the sampling criteria for small populations, using the following equation:

$$n_{ajus} = \frac{N * n}{N + n}$$

Where:

najus = Adjusted sample size

N = Population size

n = Initial sample size

Thus, considering that:

n = Number of individuals in the sample

N = Population size

$\mathbf{Z}_{\alpha}$ j2 = Critical value corresponding to the desired degree of confidence = 1.645.

p = Population proportion of individuals belonging to the category we are interested in studying = 0.5

q = Population proportion of individuals who do NOT belong to the category we are interested in studying $(q = 1 - p) = 0.5$

E = Margin of error or maximum estimation error. Identifies the maximum difference between the sample proportion and the true population proportion $(p) = 0.05$

A **sample (n)** of X families was obtained for each settlement, as described in Table 1. This sample, when adjusted by the above calculation, made it possible to draw an **adjusted sample (n j_{aus})** equal to Y families, which is what was used in the fieldwork. The results obtained for each settlement are presented in the following table, which shows the number of settlers (N), the sample (n) and the adjusted sample (n_{ajus}).

Fifty interviews were therefore carried out across all the projects, with 10 interviews in the Ilha do Coco settlement, 12 interviews in the Volta Grande settlement, 13 interviews in the Santa Emilia settlement and 15 interviews in the Martins I settlement[8] .

Table 1: Design of the sampling process

Settlement Projects - (PA)	N (population)	n (sample)	n_{ajus} (adjusted sample)
Coconut Island	29	14	9
Volta Grande	35	15	11
Santa Emilia	46	17	12
Martins I	55	18	14

Source: Prepared by the author, 2014.

In addition to the questionnaire, semi-structured interviews were conducted with the leaders of each unit, as well as with the INCRA technicians involved in the projects, in order to obtain more information about the process of formation and implementation of the settlements.

In addition, data from UAVA/INCRA, based in Barra do Garças - MT, was analyzed. This data was obtained by accessing the Information System for Agrarian Reform Projects - SIPRA. Another important source was the reports and histories held in the archive of the Araguaia Valley Advanced Unit - UAVA.

The second line of methodological analysis consisted of interpreting and classifying satellite images in relation to pasture degradation[9] in rural settlements. Using the pasture vegetation cover index, it is possible to measure the level of degradation in the settlements. In addition, by analyzing images from the initial period and the current period of each project, it was possible to compare land use

---

8 In all the settlements, an additional interview was carried out in relation to the adjusted sample (Najus).
9 The methodological process will be discussed in more detail in section IV, exemplifying the instruments and programs used, as well as the indices and images obtained.

and the development of degradation. In this way, it was possible to analyze the agricultural activities developed over the years and their level of impact on the vegetation.

CHAPTER I - THE PROCESS OF AGRICULTURAL STRUCTURE IN BRAZIL AND THE POLICY OF AGRICULTURAL REFORM: TRACKS AND CHALLENGES

This chapter will look at the historical accumulation of land ownership, as well as the mechanisms for its concentration. It will also look at the development of Brazilian agriculture, based on economic development and the modernization of this activity. Subsequently, work will be done on the country's settlement policy, which, due to systemic flaws and political fragility, is unable to implement effective agrarian reform throughout the country. Finally, the viability of rural settlements will be presented, which, even in the midst of structural difficulties, are alternatives for breaking up the country's concentrated land structure and a way for the less favored population to access the factor of production, land.

1.1 The Historical Accumulation of the Brazilian Land Structure

Rural settlements were created as a form of direct state intervention in an attempt to give rural workers access to land. According to Bergamasco & Norder (1996 apud Machado et al., 2009, p. 130), through government policies aimed at reorganizing land use, new agricultural production units are created for landless rural workers or those with little land.

Rural settlements, more than a "government policy", are the expression of the process of social organization and pressure for agrarian reform, which, however, has not yet happened. According to the 2006 Agricultural Census, 15.6% of rural establishments with more than four fiscal modules represent 75.7% of the total area occupied by agricultural establishments in Brazil; while 84.4% of the country's establishments with up to four fiscal modules occupy only 24.3% of this total area. Thus, land concentration has been a problem for Brazilian society for a long time.

Brazil's agrarian structure has been continuously designed over the centuries to favor the concentration of land. This process began in the colonial period, with the cessation of the sesmarias to the subjects of the Portuguese crown - extensive strips of land with the aim of exploiting the resources of the then colony. In the 19th century, in 1850, Law No. 601, known as the Land Law, was instituted. This law guaranteed the legality of the right to private ownership of land, in other words, land became a commodity and only those with capital could own it. As a result, the concentration of the agrarian structure worsened, so the Land Law was an important mechanism created by the rural aristocracy of the time to maintain its *status quo in* relation to land ownership. This law allowed only those who owned capital to have the opportunity to acquire land and, in addition, made it possible for farmers to sell labor to those who had no other means of living than their labor power (MARTINS, 2013).

In view of the above, in the 20th century, inequalities in rural areas sparked debate about the existence of an agrarian question in Brazil, and the idea of agrarian reform gained strength. It was mainly in the 1960s that the theoretical construction of the agrarian question gained substance, and was led by three strands of thought: the Brazilian Communist Party (PCB), the progressive wing of the Catholic Church and the Economic Commission for Latin America (ECLAC).

The latter related the "inelasticity"[10] of the rural sector's food supply and the demands of the urban and industrial sectors as structural problems in the agricultural sector, so that a change in the agrarian structure would be one of the options. Within the PCB, Caio Prado Jr. was an important theoretician who treated the issue in rural areas as first and foremost a problem of labor relations. For him, the subhuman living conditions of the majority of the rural population were the main mechanism of exclusion. Therefore, before opting for a land reform, it would be necessary to reform labor relations, with the salarization of rural workers, ensuring the guarantees of their rights, as had happened with the industrial proletariat in previous decades. The Catholic Church, for its part, played an important role in discussions about the rural environment at the time. The discussions, based on social doctrine, had a strong influence on the political and social scene, being propagated mainly by bishops, pastoral letters and social encyclicals from the Vatican, which sought to apply this model of the Catholic Church to a serious and unjust agrarian reality. In this direction, the Church's position disputed space with the left, especially the PCB, since they adopted different strategies for dealing with the agrarian question. In any case, the Catholic Church had a direct influence on the organization of Brazilian rural trade unions, as well as exerting a strong influence on the conceptualization of the right to land, legitimized in its social doctrine by the principle of social function, which was later enshrined in the Land Statute in 1964 and in the Federal Constitution of 1988 (DELGADO, 2001).

Contrary to this reformist thinking, economists from the University of São Paulo (USP), led by Delfim Netto, tried to refute the hypothesis that there was an agrarian question in Brazil. In various texts published between 1962 and 1965, the economist put forward a proposal for agricultural modernization, which would dominate the agrarian debate in the following decades. He also tried to question and present statistical data that refuted the idea of the inelasticity of agriculture proposed by ECLAC, demonstrating the thesis of the functional response of agricultural supply to the pressures of demand.

---

10 The inelasticity of agriculture was the stagnation of the food supply to the pressures of growing urban and industrial demand. In other words, in Cepal's view, the food supply would not keep pace with the growth of the urban sector and, therefore, this model was outdated and should be modified.

It can be seen that these economists considered the problem of rural areas from a strictly economic perspective, not taking into account issues such as labor relations, as pointed out by Caio Prado Jr. or the ethical and social issues raised by the Catholic Church. For this group, thinking about problems in rural areas would only be necessary when agriculture was failing to fulfill its role in the context of the industrial economy. From this perspective, the role of agriculture would be to free up labour for the industrial sector, maintain food supplies, expand exports, create markets for industrial products and finance part of the capitalization of the economy. In this way, there would only be an agricultural problem or issue if some of these functions were violated, while other issues were of little relevance in defining the role of the agricultural sector in the general context of production relations in Brazil (DELGADO, 2001).

In 1964, with the advent of the military coup and the establishment of the dictatorship, the country's political agenda was reshaped. Castello Branco's government took on agrarian reform as a necessary and structural measure for the modernization of the country's agricultural sector, as well as helping to reduce the numerous social conflicts that existed in the countryside. The regime believed that large landholdings represented an obstacle to the nation's industrialization and development process and that it was therefore necessary to combat them. That year, a working group was set up with the aim of formulating a draft land statute. The working group on this statute - or Gret, as it became known - was responsible for formulating the foundations of land reform laws and national land policy strategies. However, despite the fact that the military regime enjoyed the support of the big landowners and the rural elite, the ideas contained in the draft were harshly criticized. This dispute over the ideas and conceptions contained in the pre-project drawn up by Gret intensely marked the political agrarian debate in 1964[11] . From April, when the first version of the draft was released, to its fourteenth version in November, approved by the national congress, much was discussed and modified in relation to the original idea (BRUNO, 1995).

In fact, it is important to note that the military's thinking saw developmentalist agrarian reform as an alternative for boosting the process of incorporating capitalism into the economy. This was due to the possibility of eliminating unproductive large estates, ensuring a domestic consumer market in rural areas, low-cost food production and freeing up labor for urban areas. However, the confrontation with various opposition groups caused the policy to lose momentum and, in the end, the Land Statute was approved, torn apart from its original proposal.

It is worth noting that, even against this backdrop of political and ideological dispute, there were advances with the creation of the statute. Perhaps the greatest of all was the use of land conditional

---

[11] For more information on the power struggles over the formulation of the Land Statute see: BRUNO, R. Estatuto da Terra: entre a conciliaçao e o confronto. 1995.

on its social function, in other words, the break with the notion of private property being unenforceable and, instead, the notion of property being conditional on its use, which was a breakthrough for rural workers. From the 1964 Land Statute onwards, property fully performs its social function when it maintains high levels of productivity, conserves natural resources, creates fair working relationships and favors the well-being of owners and workers (BRUNO, 1995).

In 1967, Delfim Netto took over the Ministry of Finance and began an intense process of investment in the agricultural sector, which would later be known as the "conservative modernization of agriculture". By conservative modernization, we will use the conceptualization of Domingues (2002), who works from the point of view of the refusal of fundamental changes in the structure of land ownership, as well as the maintenance of control over the rural workforce by the large landowners. In this type of conception, the agrarian elite leads the processes of change and modernization, taking into account the interests of the landowners, "forming a 'collective subjectivity' centred on a transformist bloc, cautious and authoritarian in its perspectives and strategies" (DOMINGUES, 2002, p. 461).

This process of agricultural modernization took place thanks to the large financial contribution made by the National Rural Credit System. The ideas of the agricultural program proposed by the USP group were put into practice as public policy at the beginning of the 1960s, through measures aimed at increasing the expansion of the agricultural sector, such as increasing the technical level of labor, increasing the level of mechanization, the use of fertilizers and an efficient agricultural structure. In this way, it was possible to circumvent the need to reduce rural property in order to make it productive. With other political strategies, this period from 1965 to 1980 was the golden age of the development of capitalist agriculture, integrated with the industrial economy and the foreign sector, under the intense financial mediation of the public sector (DELGADO, 2001).

1.2   The Moving Frontier: Exclusive Modernization

It was from this process of agricultural "development" that a strong process of exclusion and abandonment of the rural environment began for those who did not have the necessary capital or advanced technology to face the frantic competition that was established in the modern agricultural enterprise. "The so-called modernization of agriculture is nothing other than the process of the capitalist transformation of agriculture, which is linked to the general transformations of the Brazilian economy" (NETO, 1985 apud TEIXEIRA, 2005).

In 1966, the national policy for the occupation of Brazilian territory was regulated under federal decree 59.428/66, which reconciled the occupation of new lands, mainly in the Midwest and North regions, with the expansion of capital in the form of subsidized credits (ORLANDI & LIMA,

2011). It was from this moment on that the process of occupation of the Midwest intensified, reflecting the emerging state policy of investment in agriculture. The nationalist discourses of "integrating in order not to deliver" and "land without men for men without land" were leveraged to carry out the strategy of creating new hubs for the development of timber, hydroelectric, agricultural and mining projects (SILVA & SATO, 2012).

The main mechanism for promoting the industrialization of the countryside and the capitalist occupation of the central region of Brazil was the agricultural credit policy. The granting of subsidized credit linked to a package of technology and specific practices led to the standardization of the agricultural model (MARTINE, 1991). It is, above all, with the incentive policy, allied to the technological package, that Brazilian agriculture goes from being classic and unproductive to modern, technified and highly profitable. In addition to this, the level of use of agricultural machinery, inputs, fertilizers and pesticides grew considerably compared to previous periods: for example, the number of tractors used in 1950 was 8,372 units; in 1960 it rose to 61,338 units, and in 1985 it stood at 665,280 units throughout the country. In addition, between 1965 and 1975, fertilizer consumption grew at an average rate of 60% a year, along with pesticides, which grew 25% a year (TEIXEIRA, 2005).

Several programs aimed at occupying the Midwest region were created during this period, such as POLOCENTRO (Cerrados Development Program) in 1975, PLADESCO (Midwest Economic and Social Development Plan), PRODEGRAN (Special Program for the Grande Dourados Region), PRODEPAN (Special Pantanal Development Program), all of which were developed by SUDECO (Superintendence for the Development of the Brazilian Midwest). PRODOESTE (Midwest Development Program) was also created by SUDAM (Superintendence for the Development of the Amazon), as well as regional development programs linked to the National Development Plan, such as PROTERRA (Program for the Redistribution of Land and Stimulation of Agribusiness in the North and Northeast) and POLOAMAZÔNIA (Amazon Development Program). All these programs, in addition to the construction of federal highways such as BR-163, BR-070 and BR364, were decisive in boosting agricultural productivity and population density in the North and Central regions of the country (ORLANDI & LIMA, 2011; CUNHA, 2006; MACHADO & CEDRO, n.d.).

It is in this set of public policies aimed at the agricultural sector that the process of exclusion of part of the rural population is accentuated. This is because the focus of the credit lines was specific to certain crops and certain production practices. This limited the access of a significant range of small rural producers to the "benefits" of the modernizing policy. Heredia et al. (2010) state that between 1980 and 2000, most of the rural credit distributed in the state of Mato Grosso was earmarked for soybean cultivation, around 50 to 75% of the amount; approximately 95% of the beneficiaries of these loans were non-family farmers, i.e. it was generally large landowners who received state support. As a result, only a portion of the rural sector, business groups, rural oligarchies and foreign capital developed and modernized under state programs.

> The heterogeneous nature of Brazilian agriculture - from a technical, social and regional point of view - was preserved and even deepened in this modernization process. In a certain sense, it can be seen as a modernizing and conservative agrarian pact which, at the same time as the technical integration of industry with agriculture, also brought into its shelter the rural oligarchies linked to large landowners and commercial capital (DELGADO, 2001, p. 165).

As a result of these selective policies implemented by the "golden age of agriculture", many squatters, sharecroppers and small producers were forced to leave the countryside because they couldn't compete with the productivity gains of agricultural modernization. In addition, land began to increase in value in these regions, which intensified land speculation.

Paradoxically, the great March to the West, induced by the state with the aim of occupying the "empty spaces", also created an exodus march from the countryside: the traditional populations that didn't adapt to the development experienced by agriculture migrated to urban areas and large centers in search of a new life opportunity. According to IBGE data, the level of urbanization from the 1980s to the 2000s was very high, ranging from 68% in 1980 to 82% in 2000 (DOMINGUES, 2002). The estimated rural exodus is around 30 million people across the country. In addition to the changes in working relationships in the countryside, wage-earning has intensified as a form of rural work, with many of these wage-earners living in urban areas (the so-called *Boias-frias*) (MARTINE, 1991).

> If it's true that fifty years ago, rural workers suffered from easily curable ailments, today they die from truck crashes or poisoning by poisons. If their houses used to have dirt floors, today they are made of the remains of wooden crates or zinc sheets in the urban slums (NETO, 1985 apud TEIXEIRA, 2005, p. 29).

This model of modernization was extremely exclusionary for the most fragile population in rural areas, which ended up generating outbreaks of social tension, many of them due to disputes over access to land involving squatters, small landowners and landowners and land grabbers. In 1979, there were 115 outbreaks of social tension in the state of Mato Grosso alone, ranging from cases involving deaths to disputes over public land without any crimes being committed. In this context, the federal government adopted measures to mitigate and solve this problem, one of the alternatives being the creation of settlements in new rural areas to house the population in conflict. In fact, from the state's point of view, squatters and rural workers in search of land for work were seen as a driving force behind agrarian conflicts, as well as being associated with the "backward" agricultural model, maintaining a traditional production relationship with little chance of growth. Therefore, the negative concept of possession is contrasted with the idea of property rights and the process of modernization that agriculture was going through, which ended up justifying the various forms of violence practiced against groups of rural workers (FERREIRA et al., 1999).

For this reason, the 1980s saw the implementation of rural settlements across the country, albeit slowly and irregularly, even though most of them were created to appease localized conflicts, without a real structural policy (BERGAMASCO, 1997). In many cases, the agencies responsible only regularized the squatters, and the expropriation actions took place in the sense of land regularization, i.e. in most cases the beneficiaries of the projects already occupied the plots even before the settlements were created.

It was therefore after Brazil's re-democratization that agrarian reform entered the political agenda, with the creation of the First National Plan for Agrarian Reform - PNRA, which proposed the creation and settlement of 1,400,000 more families in rural areas. However, the actual figure for the period of the 1st PNRA was just over 125,000 families (ROCHA, 2013). Thus, throughout all the other governments that followed this period, Brazil's land policy was always sidelined and treated as non-essential for the country's development.

1.3 Rural Settlement Policy or Land Reform Policy

According to INCRA (2014), from a formal point of view, agrarian reform would be a set of measures aimed at promoting a better distribution of land, through changes in the regime of its possession and use, in order to meet the principles of social justice, sustainable rural development and increased production. From this perspective, agrarian reform should lead to the deconcentration and democratization of the land structure, the production of basic foodstuffs, the generation of jobs and income, the fight against hunger and poverty, the diversification of commerce and services in rural areas, the internalization of basic public services, the reduction of rural-urban migration, the democratization of power structures and the promotion of citizenship and social justice.

However, the agrarian reform policy, in the form in which it has been implemented in recent decades, has proved ineffective in terms of democratizing access to land for rural workers, so that the actions carried out by governments cannot be treated as a broad public policy that seeks equity in the land environment. This is because, when we look at the government's actions regarding the process of land redevelopment, we don't see any joint action to this end. Authors such as Cunha et al. (2005) and Moreira et al. (2003) work from the perspective of the existence of a settlement policy to the detriment of an agrarian reform policy, since, in recent decades, the state has been much more concerned with resolving land conflicts in rural areas - the result of pressure exerted by workers organized in social movements such as the MST (Landless Rural Workers' Movement), and the trade union movement through CONTAG (Confederation of Agricultural Workers) and its federations - than with implementing a previously established policy.

This notion of the government simply reacting to the actions of organized groups in rural areas with regard to access to land is also identified by Heredia et al. (2002), who analyse the emergence of territorial *patches as a* result of the settlement policy. For the authors, the formation of patches is an advance, albeit timid, on the isolated process of creating rural settlements adopted by the federal government. The expropriation actions that led to settlements throughout the country during the 1990s took place mainly in regions where there was an organized rural workers' movement. The perception of the success of the path adopted encouraged workers to follow suit, putting pressure and creating more settlements. Hence the emergence of reformed areas. This close relationship between the actions of workers' movements and expropriations, with the consequent creation of settlement projects, can be seen when analyzing that 96% of the settlements surveyed were the result of conflict situations. In 89% of the cases, the initiative to expropriate came from the workers themselves, while only 10% of the settlements were created by INCRA. Actions such as land occupation were adopted by 64% of the current beneficiaries of agrarian reform, while 29% of the settlements came about as a result of resistance to the land on the part of their workers, thus showing the reactivity of the state to pressure from rural movements.

Another factor that contributes to reinforcing the idea of the absence of a national land reform policy is the logic of expropriation to the detriment of the installation of basic infrastructure. This factor is evident in Sauer's (2005) article on the quality of rural settlements, published in 2005. In the text, the author identifies a logic of favoring investments in land acquisition, as opposed to investments in actions to improve the quality of life or boost the economic development of the settlers. The prioritization of land collection is due to a logic of meeting targets, in other words: priority is given to settling families on the plots, as this generates visibility that can be expressed in numbers.

This logic of expropriation becomes clear when you realize that, in the Northern region, 49.3% of settlement projects come from expropriation, benefiting a total of 32.7% of families. If we compare this to the general index, the average number of areas expropriated is 43.1%, and the number of families settled is 27.5%. In addition, the northern region has the worst percentage in terms of quality of life in the settlements, revealing problems such as the lack of roads, electricity, health services and others. In this way, it is clear that the government's choice to expropriate areas with a larger stock of land - and consequently lower land prices - results in public spending with a greater return for the families benefiting. This logic of concentrating action on land acquisition, without investing in minimum infrastructure, penalizes families and prevents important advances in the process of democratizing access to land.

In this sense, there are many consequences of this deficit-ridden land reform policy. The most obvious is the lack of infrastructure generated by the mismatch between the government's specific actions and the needs faced by the families benefiting from these projects. In the study carried out by Guanziroli et al. (2001) on the causes of evasion in rural settlements, the issue of the PA's infrastructure was an important factor in the propensity for settlers to leave. Health services were offered in only 50.8% of the settlements surveyed[12] , while 40.7% of the settlements do not have any kind of medical care, not even first aid, which forces the settlers to travel to the municipalities for this kind of treatment (which is usually several kilometers from the settlements).

Another problem identified in the same study was the use of paternalistic practices, which contribute to strengthening the bond of dependence between these families and the local authorities. In some cases, such as Ouro Verde (PR) and Vale do Pajeù (PE), the doctor only sees patients who vote in the municipality. Added to this are the widespread diseases in the settlements analyzed, such as malaria, leprosy, dengue fever, schistosomiasis, as well as "hunger diseases" (tuberculosis, diarrhea, anemia, malnutrition, etc.), the result of *poor* nutrition and precarious living conditions (GUANZIROLI et al., 2001).

This situation worsens when we look at other factors, such as education, for example. 88.1% of the settlements have schools, but the facilities are poor, with poor lighting and sporadic maintenance. In general, school supplies are lacking and lunch is not provided in most schools. In 61% of them, classes are offered up to the 5th grade, but almost always in a multigrade system, in which a single teacher teaches three or four different grades at the same time, in the same classroom. Faced with this precarious situation, and the difficulty of transportation to nearby municipalities, many settler parents take their school-age children to the cities in order to guarantee their studies (GUANZIROLI et al., 2001).

When discussing the causes of dropout, Guanziroli et al. (2001) draw attention to the precarious conditions in schools, the irregularity of basic medical care, the poor upkeep of roads and the lack of transportation and electricity, factors which have generated a dropout rate of 29.7% throughout Brazil, with this figure varying from region to region, with the North having the highest rate (41.8%) and the Southeast the lowest (12.1%).

Still on the structural issue of the settlements, the difficulty in obtaining credit is a critical point in government policy. INCRA has a credit system for beneficiaries which is divided into different

---

12 This includes everything from health centers with doctors to the presence of health workers in each project.

modalities such as development, food and housing. The first two are generally made available at the beginning of the settlement for emergency food needs and to start planting crops on the new land. Housing loans are granted for the construction of houses with the minimum infrastructure required for decent housing. However, there are recurrent cases of delays in the payment of these incentives, which are conditioning factors for the well-being and development of the settled families.

In the article *An analysis of the regional impacts of agrarian reform in Brazil* (2013), Heredia et al. point out that the settlements surveyed were delayed by around four years for development credit and five years for housing credit, from the time the families entered the project areas, which shows how disorganized the settlement policy is in Brazil. The guarantee of land, pure and simple, is not a certainty of emancipation for formerly excluded workers, but rather a possibility, the realization of which requires the help of the state in terms of staying on the land obtained.

Because of the problems in the national agrarian reform policy, referred to here as the settlement policy, the search for new mechanisms to transform the difficulties in applying this proposal is becoming increasingly present in the debate. Abramovay (2005), for example, addresses a new perspective for this policy in *Um novo contrato para a política de assentamentos (A new contract for settlement policy)*, in which he works on the idea of creating evaluation and collection models for settlement projects. For the author, there are no comprehensive studies indicating a relationship between public spending on land acquisition, the implantation of families on plots and the release of credits, as well as all the infrastructure in the settlements (such as roads, electricity, artesian wells) and the development or evolution of these families on the land they have been given. In other words, there are no methods for evaluating the government's spending on creating the projects and their success in terms of the social and economic well-being of the families after they have been installed. In view of this, there are still doubts today about the advantages of creating a settlement project. Many public administrators wonder whether it is really worthwhile from a social point of view to invest in agrarian reform, even in the face of countless cases of default and the deficit situation of most PAs.

For this reason, the author works with the possibility of creating accountability mechanisms for the beneficiaries of land programs, opening up a parallel between urban microcredit organizations. The latter operate from the perspective of making assets available to the poorest, but under a chain of responsibility logic, in which the borrower knows that failure to pay will lead to problems not only with the agency that lent him the money, but also with the community. In addition, the credit agent also has a burden of responsibility: lending the money, monitoring and following up on the families' activities. This combination of commitment between public bodies, in order to respond to the need for credit and the assimilation of the beneficiary families, gives the idea of punishment for nonpayment; in other words, it is in the unity between the presence of the state and economic rationality that the success of these ventures lies. The essential thing is to provide incentives with rules internalized by the actors that allow them to be used properly (ABRAMOVAY, 2001).

The Brazilian process of creating rural settlements has never been based on a culture of evaluation: the state, as the managing body, has never supervised or held the beneficiaries of this policy accountable for the benefits they have received. Thus, there was no perception that the land was an advance from the state to these families and that, therefore, the resources should be used in a satisfactory and productive way. Experiences of land credit with families informed of the need for a counterpart in relation to receiving the land have produced interesting incentives to improve their productive capacity. However, what currently exists in terms of state evaluation only translates into numbers of families settled. The quantity to be publicized outweighs the quality of life of these people in rural areas. This is compounded by the fact that rural workers are certain that they will gain land without having to pay for it, leading them to believe that the plot is an asset rather than a productive base. This divergent logic must be changed to a new contractualization of settlement projects, in which their implementation undergoes analysis, so that the products of this evaluation have consequences for their actors (ABRAMOVAY, 2001).

In view of the above, it is difficult to call Brazil's attempt to reorganize land as an agrarian reform policy. The government has been unable to slow down the effects of the agricultural development model, since more than 500,000 small landowners were expelled from the countryside in the last decade of the 20th century alone. Also, as seen above, the creation of settlement projects is more a reaction to pressure from organized movements than necessarily a conception of the model.

This scenario imposes on us the certainty that what is underway in the country is not effectively an agrarian reform policy, not in the classical mold and certainly not in the revolutionary one. Among the factors that allow us to infer this assertion are: the settlement policy, which is fundamentally focused on land regularization and resolving agrarian conflicts; the continued expulsion of rural workers from the agricultural sector, which in recent decades has been greater than the number of families settled by governments; the lack of minimum infrastructure and the inhospitable nature of the regions where the settlement projects are located, which have turned into failures in many government initiatives; and, finally, the use of land purchase mechanisms to the detriment of expropriation processes, which favor speculative movements and increase the value of properties in all regions. The need for agrarian reform will be proven throughout this work, but the current models need to be rethought and debated (MATTEI, 2013).

1.4 Rural Settlements: A Viable Alternative?

Brazil's agrarian reform policy uses the term "settlement" to refer to a previously defined area in which a specific population will settle. It is therefore a transformation of the physical environment with the aim of agricultural exploitation. In addition, this specific nomenclature refers to the maintenance of workers in agricultural activity in a broader sense, thus involving the improvement of conditions for land use, the encouragement of social organization and the strengthening of family farming (ESQUERDO & BERGAMASCO, s/d).

In general terms, in Brazil there are 924,000 families[13] settled in 8,763 settlements, covering a total of 85.8 million hectares (ESQUERDO & BERGAMASCO, n.d.). This shows that the settlement projects are a vast "laboratory" of different social experiences. In a study carried out to analyze the impacts of settlements in six states of the federation, Medeiros & Leite (1997) highlighted some points: the population settled in many places is a significant part of the rural population of the municipality, indicating that these settlements are places for families to remain and work in rural areas; the PA's represented greater stability for families; in several states the population over 60 made up a large portion of the beneficiaries, which allows us to gauge that the settlements are fixers of the older population, which would probably be out of the labor market; Finally, although structural deficiencies are a limiting factor, there has been an improvement in the families' living conditions (KAGEYAMA et al., 2006).

In the study carried out by Heredia et al. (2002), a large part of the settled population already lived in rural areas, with around 80% of the sample coming from the municipality or from municipalities neighboring the settlement, and 94% of them already having experience in rural areas. When analyzing their work prior to arriving at the settlement, it was found that 75% of the workers were engaged in agricultural activity, either as permanent or temporary rural wage earners, or as squatters, or as partners, or as tenants, and so on. It is therefore possible to state that the rural settlements have enabled access to land ownership for a historically excluded population. And even if this population maintained some kind of insertion in the labor market, it did so in precarious and unstable conditions. The settlements made it possible for these people to have a place that was a source of work and conditions for social reproduction.

Another important issue regarding the importance of settlements in their regions is the role they play in diversifying the agricultural products available. In other words, these sites promote a productive reconversion in regions that were previously monopolized by monoculture or extensive livestock farming, which leads to an increase in the availability of products that were previously secondary on local agendas. For example, there has been an increase in the production of

---

13 If we consider an average number of four people per family, in absolute numbers of settled people we would have close to four million people, which is not insignificant, but far short of the rural dimension that Brazil has.

pineapples, oranges, milk, passion fruit, corn, rice, eggs, tobacco, as well as manioc flour and manioc itself. In addition to plant production, there has been an increase in cattle, pig, goat and sheep farming, which has allowed regional markets to offer a greater diversity of products at better prices thanks to increased competition. It also allows beneficiary families to strike a balance between production for sale and production for subsistence, qualitatively improving their diet and social well-being in general (HEREDIA et al., 2002).

In terms of Total Production Value (TPV)[14] , establishments benefiting from agrarian reform averaged R$15,800 per year in 2006. This is equivalent to 52.7 minimum wages at the time (R$300) per year, or 4.4 MW per month. Compared to the average family unit, which had an average annual VTP of 14,000, the settlements have a better average annual income. Similarly, the values added by the agro-industry in agrarian reform establishments had a significant share in terms of the values added by the agro-industry in all agricultural establishments in Brazil. Around 13% of the total value (R$376.3 million) came from this sector, and this share on average was higher than any other productive unit in the country (MARQUES et al., 2012).

In a study on the *socio-economic impacts of rural settlements in Mato Grosso*, Fernàndez & Ferreira (2004) state that "rural settlements have been characterized as a factor in attracting populations to new municipalities with low population density", especially in recent frontier regions. Another important factor is that in regions lacking policies to promote socio-economic development, settlements present themselves as an alternative for transferring public resources, building basic infrastructure, increasing agricultural production, and attracting populations to boost the internal market. On the other hand, what was striking about this study was that of the five settlements analyzed, four had a Gross Production Value (GPV) higher than the GPV of the family establishments in the municipality where they are located, revealing production potential in the face of the reality of local family farmers. In addition, the capacity to generate income was significant: analyzing the Average Net Family Income - RMFL[15] of the settlers, we arrived at a figure of R$339.00, which in 1997 was equivalent to 2.8 minimum wages, and was therefore higher than the exogenous poverty line, which defined two minimum wages as the ceiling. Similarly, the areas set aside for rural settlements have brought about significant changes in local development. In regions of "recent occupation", the change in the land fabric was significant, while in places of "consolidated occupation" and "transition", the greatest contribution was in terms of economic

---

14 Total Production Value includes the production, marketed or not, of animals, poultry, the sale of humus, manure and fish, crops, horticulture, floriculture, forestry, plant extraction and the added value of agro-industry, which is the total value of production minus the value of the raw materials used (MARQUES et al., 2012).
15 The RMFL is calculated from the Average Gross Family Income - RMFB, which includes the total income of the plot plus other forms of income such as retirement and sales of services, minus production costs (inputs, machine rental, freight, energy, hiring labor).

growth and alternatives for employment and income, generally presenting a capacity to resolve rural conflicts resulting from the struggle for land. In general, the social indicators show that there has been an improvement in the living conditions of the settled families, compared to the situation that preceded the creation of the projects, reflected in the increase in family consumption, especially in areas with better productive and social infrastructures (FERNANDEZ & FERREIRA, 2004).

Brazil's rural settlements represent a new form of social production, from the point of view of the families who benefit from them, who have a new dynamic in terms of working time and the performance of activities that were previously non-existent in their social relations. Thus, land ownership allows rural workers to transform their interactions and attributions in the community, becoming more active social subjects. This points to a change in the balance of power in society, since previously the landowners were the ones in control, through partnerships, leases and, above all, wage labor; with this new perspective, settled farmers now have a new weight in the balance of power (BERGAMASCO, 1997).

In the light of the above, it can be seen that the settlements are important mechanisms for social transformation and an important instrument for access to land, despite the precariousness of the infrastructure and state support. It is important to point out that the poor rural population has *been* abandoned by the public authorities, due to the precarious conditions of the infrastructure in the settlements, as explained in the previous sub-item; however, even so, these projects represent improvements in the lives of these families. It is clear, therefore, that this policy is necessary: in addition to mitigating situations of extreme poverty, the settlements have to socially emancipate rural workers, and to do this it is necessary to identify and correct the errors that exist and to maximize the benefits, so that the settlement policy stops being sectorized and fragile and becomes truly an agrarian reform policy. Rural settlements in Brazil generally have serious structural problems which often make it impossible for the families benefiting from this policy to reproduce socially. In a study of settlements in the state of Tocantins, the main causes of evasion in the settlements were associated with a lack of financial resources, problems with the infrastructure of the projects and not coming from a rural background or having any experience of the countryside (RIBEIRO, 2009).

Studies such as Sparovek (2003), Guanziroli et al. (2001) and Marques et al. (2012) point out the problems that exist in agrarian reform settlement projects. The inefficiency of the state in managing and overseeing these projects, coupled with late financial aid, creates a process of exclusion for families who are less able to adapt, either because they are inexperienced in rural areas or because they lack the resources to survive. The lack of collective infrastructure, such as schools, health centers, transportation and roads, are also factors that limit the permanence of settled families. The average dropout rate in Brazil is 29.7%, with the North and Midwest regions having the highest dropout rates, at 41.8% and 40.4% respectively. The Southeast and Northeast regions have the lowest dropout rates, with 12.1% and 15.1%, respectively (GUANZIROLI et al., 2001).

It is therefore clear that the settlements have serious problems and that these problems lead a significant number of beneficiaries to abandon their areas. On the other hand, there is a significant - and even larger - number of settlers who remain in the settlement projects, despite experiencing all these problems. The reasons why these families, beneficiaries of the settlement policy, resist are little elucidated by science, and are the subject of this paper, which seeks to characterize the factors that make it possible for beneficiaries to remain in the settlement.

CHAPTER II - LAND OCCUPATION AND AGRICULTURAL COLONIZATION IN THE STATE OF MATO GROSSO - MT

2.1 The Land Question

The historical process of occupying the state of Mato Grosso began with the bandeirantes who discovered gold mines on the Coxipó and Cuiabà rivers in the 18th century. The discovery of the mines forced the Portuguese Crown to create the Captaincy of Mato Grosso, in order to occupy and guarantee the legitimacy of the territory as being Portuguese. This was necessary because of the Treaty of Madrid, enacted in 1750, which guaranteed possession of the land by use.

However, at the end of the 18th century, there was a decline in gold mining, which meant that prospectors, large landowners and merchants had to diversify their productive activities, leading many to request sesmarias from the government in order to expand their activities and their territorial extensions. For this reason, the process of land concentration began to intensify in the region. In 1889, the year of the proclamation of the Republic, Mato Grosso's economy was basically based on extensive cattle ranching, sugar cane monoculture and extractivism. This type of economy, combined with the low population density, favored the concentration of land, income and power in the hands of rural elites. In the northern part of the state, there were rubber tappers and mill owners, and in the south, the region now occupied by Mato Grosso do Sul, there were cattle ranchers and mate producers. According to Alves (2009), it was from this territorial and economic configuration that Mato Grosso do Sul's coronelismo emerged.

A complex "legal-political apparatus" was created to mediate and legitimize interests, mainly of the rural elites, in the process of acquisition of public lands by private initiative in the state. The first land law in the state of Mato Grosso was created in 1892: Law No. 20/1982, which guaranteed the regularization of consolidated occupations, sesmarias and possessions until 1889, thus extending the time limit set by the Land Law of 1850, which was only until 1854. It also guaranteed the preferential right to buy vacant land under private control whose titles did not meet the requirements for legitimization. Thus, as most occupations took place in large areas destined for livestock, mining or agriculture, what the law essentially guaranteed was the right of large landowners to buy public land, as they were the only ones with the resources to acquire it (MORENO, 1999).

The instruments created by legislators to guarantee the right to property to only a specific portion of Brazilian society have acquired various facets and models. In a study by Moreno (1999) on the impact that the Land Law of 1850 had on land occupation in the country, there is a similarity of treatment in the various units of the federation. This symmetry basically involves three aspects: the first concerns the adaptation of state laws to the interests of squatters. In all the states, the deadlines for legitimizing possessions were changed from 1854 to 1889, and in many places this deadline was extended to 1930; the second aspect was the process of private occupation of public lands, which continued to take place in all regions and states; and, finally, this process of concentration of vacant lands in the hands of private sectors was also strongly linked to the socio-political phenomenon that occurred strongly in Brazil during the First Republic: coronelismo.

Another serious problem was the series of abuses that occurred in the attempt to occupy the Midwest region, especially Mato Grosso. The acquisition of vacant land was rampant, and the colonization concessions had minimal requirements, which were still not met or monitored. Land brokers had full powers to negotiate the regularization of large areas without the intervention of state bodies[16] In this context, companies and large landowners imposed strong pressure on the brokers, so that between 1982 and 1930, many state lands went into private hands through regularization, concession or legitimization of possessions obtained by such intercessors. Added to this was the creation of the National Soil Population Service by the federal government in 1907, encouraging migration to this region, both by foreigners and Brazilians from other areas. It was against this backdrop that the colonization policy gained momentum, providing large tracts of land to companies willing to colonize a small part of their property. However, this colonization was almost never implemented: from 1899 to 1924, 152 areas of different sizes were granted in the state of Mato Grosso, and only 152 plots were colonized, covering a total area of 4,814 hectares. One of the best known concessions was that of the Erva Mate Laranjeira company, which exploited an area of more than three million hectares in the south of Mato Grosso, without fulfilling the obligation to colonize part of its "parcel" of territory (ALVES, 2009).

During the first Vargas government, between 1930 and 1947, in order to avoid the coffee overproduction crisis, the aim was to diversify agricultural production. At the same time, the occupation of Brazil's demographic voids became a tactic for guaranteeing national sovereignty, driven by a policy of occupying the central region of the country, known as the "march to the west" (ALVES, 2009). This policy, with its strong progressive ideal of industrialization, led to an intensification of conflicts over land in the region, since land was no longer seen in terms of its use value and began to be seen as a commodity - a process that resulted in the expulsion of many rural workers from their territories, increasing the already concentrated land structure in Mato Grosso (SILVA & SATO, 2012).

---

[16] The mediators were surveyors registered with the state land agency and appointed by the governor to demarcate private properties in the process of land regularization in the state. In this sense, these mediators had full powers over the land demarcation process and technical knowledge of the sales process, which made it possible to circumvent the law, since these mediators were not subject to any kind of supervision by the state government.

The Roncador Xingu Expedition was launched in 1940, and its main objective was the exploration and colonization of "unknown" regions[17] . Operating in regions such as the Araguaia Valley and the Xingu Valley, where there were several indigenous populations, this expedition began colonizing the area, establishing some places for contact with these native populations. These base posts later gave rise to municipalities such as Nova Xavantina, created on the banks of the Rio das Mortes. In 1943, the expedition gave rise to the Fundaçao Brasil Central - FBC, which continued to carry out the settlement policy in the central region. In 1966, the FBC was abolished, giving rise to SUDECO - Superintendência do Desenvolvimento do Centro-Oeste, an organization created to support the military regime's development policy (MORENO, 1999).

With the advent of the military dictatorship in Brazil in 1964, there was a significant intensification of the occupation and exploitation of the central region of the country. The various colonization programs promoted by the military regime aimed not only to occupy less populated regions, but also to boost agriculture, which until then had been characterized by low productivity and on a large scale, i.e. producing little on large tracts of land. In an attempt to modernize agriculture, the state of Mato Grosso received federal programs from the National Integration Program - PIN, in 1970. These programs primarily served to finance access to land for large economic groups (ALVES, 2009).

Furthermore, as in the rest of the country, the process of agricultural modernization proved to be extremely exclusionary in relation to small producers, as it implemented a package of technologies directly linked to large landowners focused on export agriculture and industrial processing. In short, what has happened in recent decades with large-scale agriculture has been an association with a single production model, characteristic of Brazil, which consisted of the use of large tracts of land, with a high level of agricultural mechanization and inputs, achieving high productivity. It went from a traditional and inefficient form of agriculture to the model of an agricultural company, in fact, associated with the international market.

The Midwest region of Brazil, especially Mato Grosso, has always been associated with strong emigration, which intensified the occupation of the state throughout the 20th century. This frontier movement is historically about the displacement of the workforce from central areas, where capital accumulation takes place, to peripheral areas. In this sense, it is possible to refer to the state of Mato Grosso as an agricultural frontier region:

> From the moment it set foot on the western plateaus, the pioneering wave continued on its indefatigable march, shaken by economic crises and sometimes accelerated by the play of

---

17 It cannot be said that these regions were unknown or unpopulated, as there were many indigenous people living there. However, according to the government, these traditional populations were not really part of the Brazilian state.

The occupation of the Brazilian west began in São Paulo, driven by coffee growing, around 1870 to 1880, reaching the western plateaus. São Paulo's progression lasted until around 1920, when the borders with the state of Mato Grosso were reached. In reality, the background to this process of occupying the territory was the distribution of a mass of unemployed people who were in the large urban centers, as well as providing work for the northeastern migrants, in other words, a way of mitigating social conflicts that might arise in the more populated areas. Another important point is that the closure of the southern "frontier", with the labor crisis, fueled the expansion of occupation fronts in the Amazon regions. The frontier in motion is Brazil's ability to expand its production extensively due to the availability of land during periods of high demand on the world market (CASTRO et al., 1994).

For this reason, it is essential to understand the movement of the border as something other than the individual motivations of the migrants, but rather as a reflection of the economic and social structures in the countryside, based on the overall economic formation of Brazil. The border has various functions, both socially and politically, and economically. In the latter, it functions as "a kind of regulatory warehouse" for food prices. In the political sense, the border acts as an "escape valve" for social conflicts: from Getúlio Vargas through the military regime, Brazil has always used these "new zones" to put an end to the need for land in the big centers. For the state, it was preferable to look for new places to occupy land than to change the country's land structure. On the social level, in turn, it was an opportunity to recreate peasant production expelled from the more developed agricultural regions, in other words, the possibility for small expropriated producers to obtain land for their own reproduction (CASTRO et. al., 1994).

In light of the above, understanding that, historically, Mato Grosso has always been seen as a frontier region is an important step towards understanding the process of modernization that it underwent throughout the 20th century, especially in the last decades of this century, when there was a surge in population and economic growth, giving it the mark of an agricultural region, with record grain production and cattle breeding. It was from this process of agricultural development that a strong context of exclusion and abandonment of the rural environment began for those who did not have the necessary capital or advanced technology to face the competition that was established in the modern agricultural company.

2.2 Official Colonization

In addition to national territorial integration policies, the state of Mato Grosso underwent an intense process of directed colonization, mainly orchestrated by the federal government. It was from the 1970s onwards that colonization policies emerged with greater vigour. According to the National Integration Plan (Plano Nacional de Integraçao - PIN), their objectives were: to create the conditions to incorporate large swathes of land into the market economy; to move the agricultural frontier to the Amazon River; and to redirect the migration of labor from the Northeast to the Center-South towards the West. The term colonization, in its broadest sense, means the process of occupation of an area by outsiders, settlers. Historically, however, this terminology has been used to characterize the settlement of regions previously planned by private initiative or the government, and is closely related to the private appropriation of land use (CASTRO et al., 1994).

The colonization model used by the state in Mato Grosso was under the guidance of the recently created INCRA - National Institute for Colonization and Agrarian Reform. At the beginning of the 1970s, INCRA tried to implement the Integrated Colonization Projects - PICs, a rigid and bureaucratized model with very ambitious objectives. Later, the agency itself evaluated the experiences with this colonization model as negative, since its methodology, based on the Land Statute, ended up greatly increasing the cost of these projects (CASTRO et al., 1994).

Between 1980 and 1992, 59 official colonization projects were set up in Mato Grosso, which were classified in three different ways according to their implementation strategy. The Rapid Settlement Projects were set up in areas with existing infrastructure, with support from the state and municipal governments; the plots averaged 50 hectares and occupied a total area of 269,948 hectares, settling 4,525 families.

The Joint Action Projects were carried out in partnership with cooperatives: the government provided land, basic infrastructure and land titles, and the companies' counterpart was to maintain and manage the settlements.

Finally, there was the Special Settlement Project, which is the sole responsibility of INCRA and was set up to cater for people coming from regions of land conflict. Only one was set up in Mato Grosso, located in Lucas do Rio Verde, with an area of 200,000 hectares, and settling more than 972 families from Rio Grande do Sul. It was established in 1981 and became known for land buying and selling scandals (LAMERA et al., 2008).

Private colonization projects were also undertaken in the state, and were more successful than official projects, because they obliged the companies to provide minimum infrastructure, such as access roads, warehouses, schools and health posts, as well as technical assistance and credit to the settlers. In the 1970s and 1980s, 88 private colonization projects were set up by 33 companies. 3.25 million hectares were occupied by approximately 200,000 settler families, mostly from the south, with plots on average larger than 100 hectares (LAMERA et al., 2008). However, even though these projects had better results than the official policies, a large part of the areas earmarked for colonization remained in the possession of private companies, which, once they had appreciated in value, sold them on. The Integraçao Desenvolvimento e Colonizaçao (INDECO) company alone owned 907,691 hectares, with the aim of settling 3,830 families in the Alta Floresta, Paranaita and Apiacàs projects. In addition to this, there was an intense migration flow of small squatters and the needy population, who moved to more distant regions that were not assisted by the state, due to the impossibility of accessing the official programs, whose supply was less than the demand, and the impossibility of joining the private programs, due to the high cost of the plots offered by the companies. Subsequently, a large part of the conflicts and tensions over access to land involved this mass of people who were not officially incorporated into Mato Grosso's colonization policy (FERREIRA et al., 1999).

The first official colonization project in the state, coordinated by INCRA, took place in 1978, as a result of a conflict between settlers from the Nonoai and Guarita indigenous reserves in Rio Grande do Sul. The government tried to solve the problem by bringing settlers from the south to the Terranova region, with the help of the Coopercana cooperative, which originated in the south. This union between the government and the cooperatives served mainly as an instrument for directing agricultural development policies, which could suffer some kind of interference due to the intense social struggles that were taking place in the countryside, especially in the south of the country. This pact between the government and the cooperatives turned the latter into legitimizing agents of the military dictatorship's policies, making them co-participants. Six joint settlement projects were set up in the state, each with different economic and political interests and their own specificities. Table 2 below shows the projects carried out in Mato Grosso:

Chart 2: Characteristics of the colonization projects created in Mato Grosso - MT, between 1978 and 1981.

Name	Year of creation	Total area (ha)	Plot size (ha)	Partner Cooperative
Terranova	1978	450.000	100 e 50	COOPERCANA

Peixoto de Azevedo	1980	133.000	45	COTREL
Ranch	1980	16.000	250 e 74	CAMAJUL
Lucas do Rio Verde	1981	220.000	100	COOPERLUCAS
Braço do Sul	1981	115.000	100 e 50	CIRA
Carlinda	1981	96.000	50 a 500	COTIA

Source: Nédélec et al., 2005.

Thus, from 1970 onwards, there were two types of colonization in Mato Grosso: official and private. The latter was predominant in the state, and was also the choice of public bodies such as INCRA and CODEMAT, which, tasked with promoting the occupation of vacant state lands, passed them on to colonizers. These private projects were dominated by farmers from the south of Brazil, who reproduced the southern colonization pattern in Mato Grosso, based on the founding of towns, the sale of rural plots and the selection of settlers, who had to have prior capital for the venture. As a result, some of these projects later gave rise to municipalities such as Canarana, colonized by Cooperativa 31 de Março Ltda; Agua Boa and Nova Xavantina, colonized by CONAGRO, owned by pastor Norberto Schwuantes; and Vila Rica, owned by Colonizaçao Vila Rica, among others (GALVÂO, s/d).

In view of the above, it is clear that, just as the policies for occupying the Midwest had a strong impact on the formation of the state of Mato Grosso, the colonization policies, both official and private, were the foundations of the current agrarian configuration of Mato Grosso. Certainly, these programs brought a new dynamic to the state, with the incorporation of a distinct population with a different way of life from that commonly employed. This policy was also highly exclusionary because, as has been shown, many colonizers selected farmers based mainly on the availability of capital. This meant that a large number of farmers were excluded from the official colonization mechanisms, forcing them to look for other ways to occupy a plot of land in the central region of the country.

2.3 The Dynamics of Rural Settlements: a perspective from the state of Mato Grosso - MT.

Currently, in the state of Mato Grosso, there are 545 settlement projects under INCRA's jurisdiction, with 84,501 beneficiary families, occupying an area of 6,077,078 hectares[18] , as can be

---

18Available at: http://www.incra.gov.br/sites/default/files/uploads/reforma-agraria/questao-agraria/reforma-agraria/relacao_de_projetos_de_reforma_agraria.pdf

seen in figure 1.

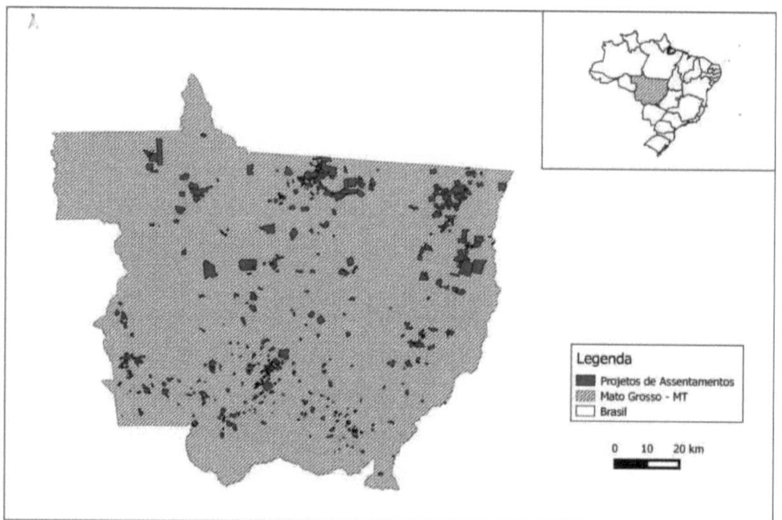

Figura 1: Settlement projects under INCRA jurisdiction in the state of Mato Grosso, MT.

Source: UAVA/INCRA, 2014.

These settlements are spread across all regions and have varying degrees of implementation, infrastructure, size and economic and social development; in other words, they are different in their formation and structure (INCRA, 2014).

The creation of rural settlements in Mato Grosso began at the end of the 1970s, still under the military regime, as a response to the social tensions in the region. In 1985, with the creation of the First National Land Reform Plan - PNRA, this reactive logic continued. This is because, even though Mato Grosso is a state known for having vacant public lands and unproductive estates, the forecast for areas to be expropriated between 1985 and 1989 was 3.5% of the areas planned for the whole country. These figures, according to INCRA-DF planning director Guilherme Muller, are due to the fact that the PNRA was not intended to distribute land, but only to regularize areas of tension and conflict. Therefore, the needs of each state were calculated based on the number of landless people and squatters (FERREIRA et al., 1999). Over the last few decades, the creation of rural settlements in Mato Grosso was most significant between 1995 and 2007, with a total of 501 new settlements, as can be seen in Figure 2.

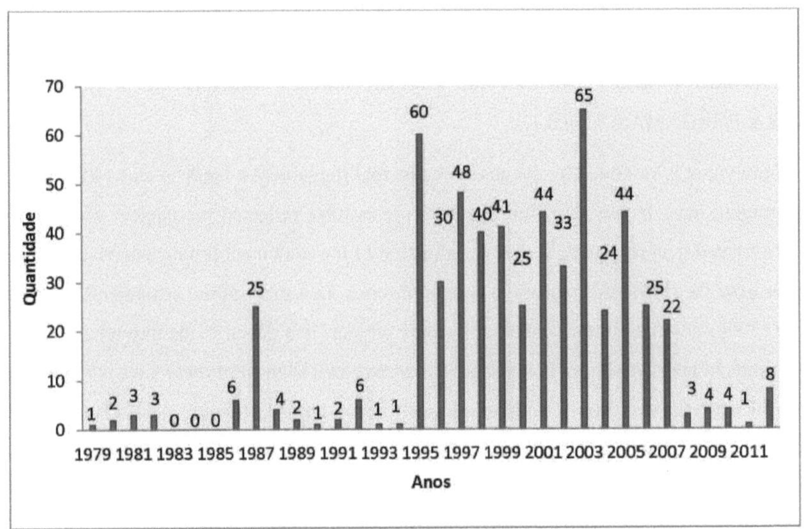

Figura 2: Rural settlements created in Mato Grosso, MT, 1979 - 2012.

Source: DATALUTA - Land Struggle Database, 2013.

The current scenario is not very favorable for the beneficiaries of the settlement policy promoted by the state. While still under the Fernando Henrique Cardoso administration, the practice of expropriating areas and collecting land from the state was replaced by incentives for direct purchase mechanisms, especially in the second term. Furthermore, under the Lula government, of the more than 36,000 families settled, only 24% were settled on land that had been expropriated or bought, with 76% being allocated to existing settlements. In reality, what we have today is a policy represented in numbers, in which new people are inserted into government programs to replace those who left, without significantly expanding access to land (ESQUERDO; BERGAMASCO, s/d).

For INCRA, Brazil's agrarian policy has improved in recent years: the number of families benefiting between 2003 and 2010 was 614,093; 3,551 settlements were created in the same period; and the area allocated to the settlement policy increased by 129%. As a result, in 2010 Brazil had 85.8 million hectares incorporated into the agrarian reform, and 8,763 federal settlements, with approximately 924,263 settled families.

On the other hand, the Pastoral Land Commission (CPT) states that 2010 was the worst year for the settlement policy, as there was a 44% reduction in the number of families settled compared to 2009, as well as a 72% reduction in the area set aside for rural settlements. This is due to the fact that INCRA's budget has been halved compared to the previous year, revealing that the policy of structural reform of the countryside is not a priority for the federal government. The settlement policy in place is not seen as a project for a nation and equitable development, but rather as a precarious settlement program that falls far short of the real demands of rural workers (ESQUERDO & BERGAMASCO, n.d.).

Evidently, a policy that is sidelined by the governments that implement it tends to show little in the way of encouraging data. In this direction, the settlements have reflected the neglect with which they have been treated over the years, so that the situation of the settlement projects in Mato Grosso does not differ from the rest of the country in structural terms. In a study of the settlements, Lamera & Figueiredo (2008) pointed out that only 7.1% of the projects had paved roads throughout, while 51% had dirt roads in poor condition. In addition, the access time to most of these sites is more than an hour from the nearest municipality, with an average travel time of 1.19 hours. These factors limit the development of settlements, since easy access to urban areas improves the marketing of production, resulting in higher incomes. More distant and remote locations often depend on middlemen to sell their produce, which reduces the real earning power of these farmers.

In an analysis of rural settlements in Mato Grosso, Alves et al. (2009) identified that, in general, the situation is precarious: in addition to serious infrastructure deficiencies, there is a strong dependence on income from outside the plots on the part of the beneficiary families. This is because the need to supplement income is fundamental to the survival of the family, which cannot develop on its property alone. In 80% of the families, the productive activity is focused exclusively on subsistence, which shows that the plots basically help to supplement income, and not as the main source of income. This external income is practiced by 37% of the beneficiaries who work inside or outside the settlements. In addition, 22% of the families receive some kind of social benefit from the federal government, showing that 59% of the settlers have access to some kind of external income (ALVES et al., 2009).

In short, studies referring to the Midwest region, specifically Mato Grosso, such as Alves et al. (2009) and Lamera et al. (2008), as well as studies such as Sparovek (2003) and Kageyama et. al. (2006), who work with national data, show that infrastructure data is precarious in the settlements: *there is* little input and support from the competent bodies, a lack of incentives and rural credits, low or sometimes non-existent technical advice and a variety of other factors that limit and impede the progress of the beneficiaries. Paradoxically, all these studies also point to an improvement in the quality of life of the beneficiaries, i.e. in general, the settlement policies improve the well-being of the settlers.

On the other hand, due to the precariousness of infrastructure and access to lines of financing and technical assistance, which characterizes the process of creating rural settlements nationwide, there is a high dropout rate in these projects. In the state of Mato Grosso, the dropout rate in rural settlements was 47.3%, higher than the national rate of 29.7% (GUANZIROLI et al., 2001).

The factors that cause evasion have been the subject of studies analyzing rural settlements, especially in the 1990s and 2000s. Based on these studies, it can be seen that evasion from rural settlements does not originate from defined factors, but is a reflection of a set of situations and problems that trigger this exit from the projects as an alternative for the families benefited. Infrastructure conditions, poor administration by public bodies, delays in the release of credit and financial support, and the lack of capital on the part of the settlers are all factors that have been identified as potential drivers of evasion in the settlements. Even if the existence of these factors is not a determining factor in the decision to leave, they do reflect the conditions and environment for this to happen.

In light of the above, it is understood that the conditions of the settlement projects in the state of Mato Grosso, and in general in Brazil, reflect the low priority the government has given to promoting a real agrarian reform policy, limiting itself to a policy of settlements, which are punctual and uncoordinated in their actions. However, in the midst of this environment, it can be seen that there is a considerable number of people who remain in the settlements, which is even higher than the number of people who leave. It is important to remember that, despite the poor structural conditions, the lack of support from the state and often the isolation of these projects, there is a land market in the settlements, which promotes mobility between people. In many cases, there is an interest in securing land ownership, and rural settlements guarantee this possibility. In this sense, evasion is expressed more by the parallel land market than by the total abandonment of plots, since at the same time as some settlers evade, new workers and land seekers arrive in the settlements.

Thus, in the midst of the precarious situation in which the settlement projects find themselves, understanding the reasons for staying is to shed light on a problem that has been hotly debated over the last few decades: evasion in rural settlements. What are the reasons for beneficiaries to remain in these projects? Studies on settlements do not address this issue, and it becomes pertinent as we realize the difficulties that exist in the projects and the existence of considerable evasion. Furthermore, to elucidate permanence is to understand what the main factors are for the policy to strengthen the beneficiary families and really promote their social and economic emancipation.

Thus, even with significant problems, rural settlements are important livelihoods. Throughout Brazil, there are 1,320,463 settled families who seek their social reproduction in rural areas (INCRA, 2014). In view of this, permanence is an important factor in settlement policy and should be better understood in order to strengthen state support for the beneficiaries of rural settlements.

CHAPTER III - RURAL SETTLEMENTS IN THE ARAGUAIA VALLEY OF MATO-GROSSENSE: THE PERMANENCE OF THE BENEFICIARIES OF THE SETTLEMENT POLICY

Understanding the reasons for remaining in rural settlements is an important point for improving public policy on land redevelopment. Thus, understanding the reasons why these settlers remain in the settlements is an important tool for advancing the construction of an effective agrarian reform policy. How and why, in the midst of so much adversity, do farmers stay in the countryside? These questions permeate the entire discussion of this work, in an attempt to identify the characteristics of these beneficiaries, which differentiate them from a significant proportion of other producers who have not endured the conditions of life in rural settlements. As explained in the previous chapters, the difficulties encountered in rural areas involve a range of factors, including the lack of a structured policy for the development of the countryside, the absence of public power and the constant lack of resources for investment. As such, dropout is a reflection of the lack of action by public authorities, coupled with the precarious infrastructure available, which creates a situation of instability for the population that lives there.

This chapter will present the history of the settlements surveyed and the results obtained from the questionnaires applied, characterizing the socio-economic profile of the beneficiaries, as well as the infrastructure characteristics of these projects. It will also present the inferences drawn from the interviews conducted with settled workers, local leaders and INCRA technicians, in an attempt to highlight the elements that allow us to understand the permanence of workers in these settlements.

3.1 History of the Rural Settlement Projects Analyzed

3.1.1 Santa Emilia Settlement Project

The Santa Emilia PA was created in 2003 by INCRA decree 117/03, as a result of the expropriation of a farm with the same name in the municipality of Barra do Garças - MT. This project is registered under the number MT0616000, and covers an area of 1,387 hectares, divided between forty-five plots ranging from 26 to 35 hectares each. The location of the settlement and its plot structure can be seen in figure 3.

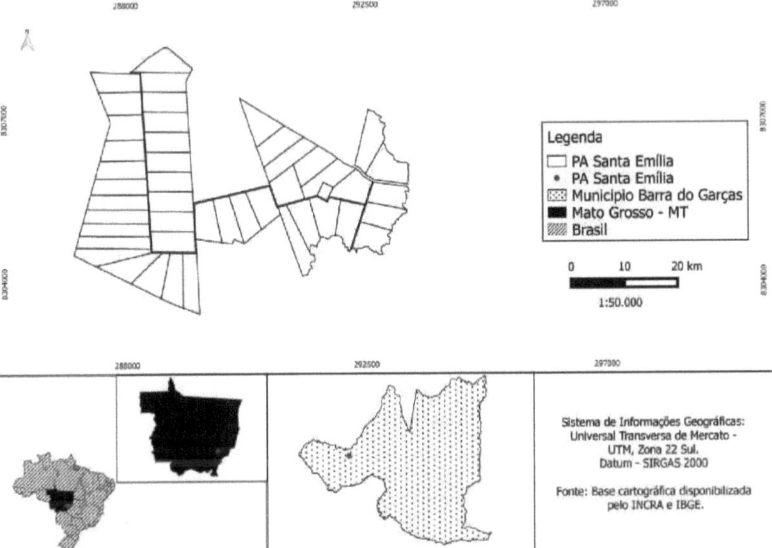

Figure 3: Location and plot structure of PA Santa Emilia, Barra do Garças, MT. Source: Prepared by the author, 2015.

Located in Barra do Garças, it is 125 km from the municipality's headquarters and 50 km from the municipality of Novo Sao Joaquim, and the roads linking the settlement to these towns are not paved. The settlement is located about 20 km from the Toricoeje district, where there is a basic urban infrastructure used by the project's beneficiaries, such as a municipal school, supermarket, public telephone, transportation to other localities and a health center.

According to the president of the Association of Small Producers of the Santa Emilia Settlement Project, the settlement was created through the organization and demands of rural workers for land. This process began at the end of the 1990s, when landless rural workers occupied a farm in the municipality of Novo Sao Joaquim, with the aim of putting pressure on public bodies to expropriate and allocate the land for agrarian reform. However, the owner of the farm challenged the occupation in court and later won the case, forcing the farmers to vacate the property. After this process, the landless workers remained near the farm, camped by the side of the road. At the beginning of the 2000s, the workers learned that Fazenda Santa Emilia, in the municipality of Barra do Garças, was in the process of being expropriated by INCRA. As a result, the campers went to the vicinity of this farm in order to put pressure on it and guarantee the occupation of the area once it had been expropriated.

In 2003, a writ of possession was issued and INCRA authorized the occupation of the area by the encamped workers, in order to wait for the demarcation of the area within the property. In the same year, the plots were demarcated and drawn for each beneficiary. According to the president of the association, this process took almost two years on the Santa Emilia farm alone, plus another two years in the previous encampment, in the municipality of Novo Sao Joaquim. During this period, there was no conflict with the landowners, not even with the police.

In a nutshell, the process of creating the Santa Emilia settlement began with an occupation carried out by rural workers on a farm in another municipality; prevented by the courts from remaining on that property, they marched to the region of the current project in an attempt to secure its expropriation, which was already in legal proceedings.

Currently, the settlement has 48% primary residents, i.e. those originally settled by INCRA. Infrastructure conditions are problematic in some respects, such as the difficulty of accessing urban centers and the lack of assistance. However, among the settlements analyzed, this was the one that showed the greatest socio-economic organization among the settled families. One of the points that reinforces this perception is the community's attempt to organize a dairy area in the settlement, so that they can sell enough to the dairy in the city of Novo Sao Joaquim. Dairy farming was repeatedly cited as the most viable option for the settlers, as it provides monthly financial income.

In view of the above, the Santa Emilia project, although it has structural limitations, enjoys active organization among the settlers. As it is the most recently created settlement among the four analyzed in this research, it also has the presence of primary beneficiaries, who were there at both times, the encampment and the struggle for land, making the group more cohesive.

3.1.2 Volta Grande Settlement Project

The Volta Grande settlement had its creation decree ratified in 1991, by INCRA Ordinance No. 070/91. However, the beneficiaries had *been* occupying the area since 1988, and were granted possession of it on September 13, 1988. The project's registration number is MT0055000 and it covers an area of 1,745 hectares, divided into thirty-five plots ranging from 30 to 82 hectares. This settlement is located in the municipality of Araguaiana, from which it is 25 km away, and is also 50 km from the municipality of Barra do Garças. It is located on the banks of the Araguaia River and most of the access to the settlement is via paved roads. Due to its proximity to the municipal center and the ease of access, there is no basic infrastructure such as a school, health center, supermarket or public telephone in the settlement.

In an interview with a settler from Volta Grande, the current president of the Rural Workers' Union (STR) in Barra do Garças, it emerged that the process of expropriation of the project began in 1982, with the occupation of the farmland by landless farmers. As a result of this dispute over the land, there were conflicts with gunmen, but no "tragedies", as the interviewee revealed. In the midst of all this, the landowner filed a legal challenge forcing the campers to leave the property. After several legal disputes, in 1988 the area was expropriated and made available for agrarian reform, which made it possible to reoccupy the area in September 1988. Figure 4 shows the location and structure of the Volta Grande Settlement Project.

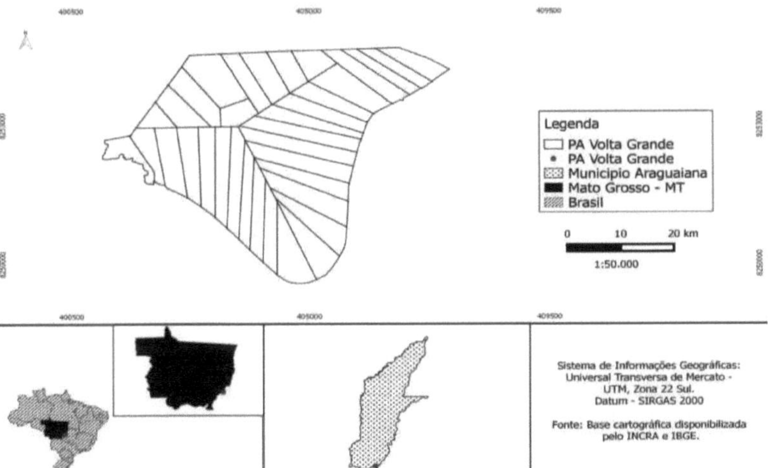

Figure 4: Location and plot structure of the Volta Grande PA, Araguaiana, MT. Source: Prepared by the author, 2015.

Due to its proximity to the municipalities of Barra do Garças and Araguaiana, as well as its privileged location on the banks of the Araguaia River, the Volta Grande settlement has peculiar characteristics in relation to other agrarian reform settlements, such as the sale of land on the banks of the Araguaia River for leisure. The use of this practice has become a form of capitalization for the beneficiaries who find themselves without resources or access to public funding. Another important factor was the identification of an inn within the settlement[19] , on plot 23, set up because of the great tourist appeal that the region offers, especially on the Araguaia River.

It is an old settlement, approximately 27 years old, and a large part of the beneficiaries already have title to their plots: 29 plots already have title. For this reason, the beneficiaries don't recognize themselves as belonging to a settlement. This is evident from the fact that in the plots visited it is common to name the place sitio or chàcara, which reflects the idea of ownership and dominion over the land. Added to this is the lack of a residents' association in the settlement. So, for the most part, what you see in the Volta Grande PA is a cluster of small properties that are not necessarily organized or linked to each other.

19 The hostel is called Cristal. And it does not belong to the beneficiary of the plot where it is located. In reality, the beneficiary parcelled out the plot and sold part of it to a third party, who used the acquired area to create this development.

The families, despite being originally part of a settlement project, behave like traditional farmers, without maintaining the sociability ties they have. A reflection of this lack of articulation between the settlers is the predominance of livestock farming, a typical activity in the region, but mainly related to large landowners. It became clear during the interviews that a significant proportion of these beneficiaries, in addition to relying on production on their plots, have an external source of income. In other words, they have jobs in the city or even in the rural sector or are retired, so that, to a large extent, they have agricultural production as a source of supplementary income.

3.1.3 Ilha do Coco Settlement Project

The Ilha do Coco settlement was granted possession on June 11, 1987, and was created by INCRA decree 691/87, under registration number MT0039000. This project covers a total area of 2,828 hectares, divided between 34 plots ranging from 25 to 99 hectares. Located 25 km from the municipality of Nova Xavantina and 150 km from the municipality of Barra do Garças, it has reasonable access conditions to the urban area, as the dirt road has good traffic conditions all the way to the municipality's headquarters. As in the Volta Grande PA, due to the ease of access to the municipality's urban area, there is no basic infrastructure in the settlement, such as a school, health center or public telephone.

The Ilha do Coco PA came about as a result of a territorial dispute between the workers who occupied the property and a land grabber who had the documents for another area, located in an indigenous reserve, but who pressured the occupying families to leave the place, claiming that he owned it. As a result of this dispute, INCRA collected and expropriated the property for agrarian reform purposes and settled the nine families who had originally occupied the property. In creating the settlement, INCRA included other workers as beneficiaries, such as those who had been evicted from the Piau gleba, another property occupied near the newly created project. In addition, as the total capacity of the settlement had not been completed, INCRA also selected rural workers from the municipality of Nova Xavantina to complete the total capacity of families benefiting from the creation of the project. The location and structure of the Ilha do Coco Settlement Project can be seen in figure 5.

Through the documentation available at the Advanced Unit of the Araguaia Valley - UAVA/INCRA, it was observed that after the settlement was created, 54 beneficiaries had already been approved to occupy the 34 plots it has. Currently, there are only 27 beneficiaries who have been approved, and another 7 families awaiting approval, so there are no abandoned plots in the settlement. Thus, of the beneficiaries who have already been through the project, added to the families who have not been regularized, there have been a total of 61 beneficiaries throughout the settlement's existence, showing a turnover or evasion rate of 79%.

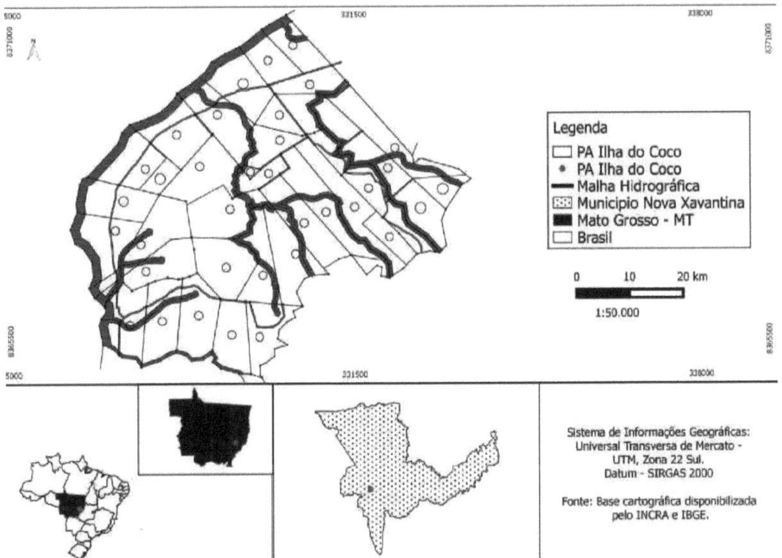

Figure 5: Location and plot structure of the Ilha do Coco PA, Nova Xavantina, MT. Source: Prepared by the author, 2015.

Another important point noted during the field research was the reconcentration of the plots among the resident families. In other words, families who own three plots, one of which belongs to the mother and the other two to her children, have transformed everything into a single property, with a single house, a single infrastructure, such as a corral and machinery, in order to have more pasture and more profitable use of the area, facilitating management and optimizing work. As in the Volta Grande PA, the families in the Ilha do Coco PA also lacked unity, presenting themselves more as an agglomeration of small landowners than as a rural settlement resulting from the struggle for access to land. This becomes clear when you realize that the settlement's Residents' Association no longer plays its role as it used to, and has no organized structure, such as an elected board of directors, or frequency of meetings.

3.1.4 Martins I Settlement Project

The Martins I settlement project was created in 1987 through the collection of public land by INCRA, as this area previously belonged to the Union. The project is registered under the number MT0067000 and has an area of 3,848 hectares, divided between 54 plots ranging from 62 to 78 hectares. Located 130 km from the municipality of Agua Boa, it is the project with the longest distance to the municipal center of all those surveyed. Access to the project is precarious, as the dirt road becomes impassable for smaller vehicles in rainy weather.

The creation of PA Martins I arose from the need to assist 51 families evicted by court order from the then Gleba Volta Grande, in the municipality of Barra do Garças, who were camped on MT-100. The area of the Martins gleba was taken over in the name of the Union, according to INCRA-DP-014/87, with the aim of promoting the settlement of landless workers. Of the 51 families evicted from Gleba Volta Grande, 31 were transferred from the camp to the settlement. The total number of settlers was later supplemented by a further 23 families selected by INCRA from among the landless rural workers of Agua Boa, bringing the total number of families in the project to 54. The location and structure of the Martins I Settlement Project can be seen in Figure 6.

[a]In terms of infrastructure, this project is better structured than the others surveyed because, in addition to a school in the area up to the 3rd grade of secondary school, it also has a health center, located inside the school, where a nursing technician and a health agent work, serving the public not only of the settlement, but of the entire region in which it is located. There is also a public telephone located near the school and a community milk cooling tank, which is available to everyone in the settlement and has its milk collected weekly by a dairy in the city of Agua Boa. All this support, according to the residents, comes from the town hall, which provides resources for the rural school, as well as guaranteeing the rent for the milk tank and the salaries of the nursing technician and the health worker.

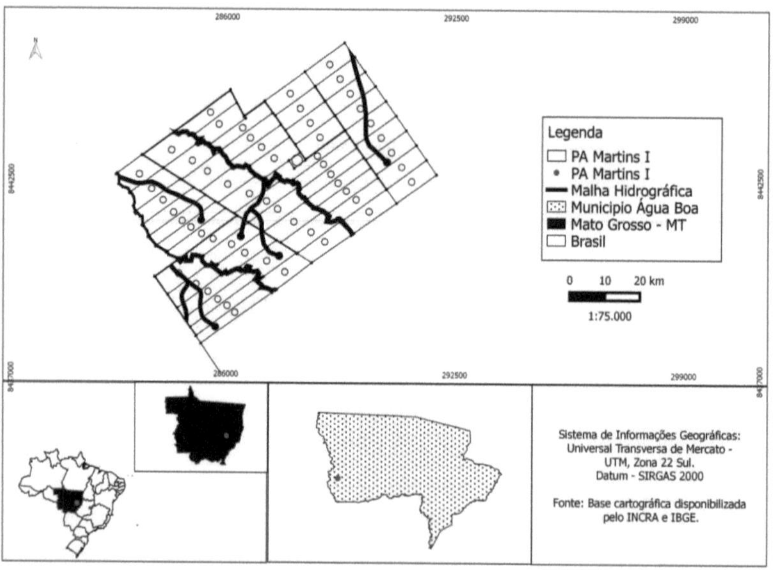

Figure 6: Location and plot structure of PA Martins I, Agua Boa - MT.

Source: Prepared by the author, 2015.

It became clear during the field research that the infrastructure provided by the municipality provides better conditions for the settled workers and their families. This is confirmed by the greater presence of young people aged 0 to 18 in the project, approximately 38.5% of the total number of people living in the settlement - a rate that is even higher than the percentage of young people in the other settlements surveyed. In a way, the existence of a school up to the third year of secondary education has allowed young people to remain in rural areas, not forcing them to migrate to urban areas in search of an education. The existence of a milk cooling tank enabled the settlers to stock up on milk during the week, making it possible for the dairy in the municipality of Agua Boa to collect it.

During the field research, it was found that there was an elected president of the project's Residents' Association, but just like in the Ilha do Coco PA, there was no organization or frequency of meetings. The president was more of an interlocutor for the settlers on issues to be dealt with by INCRA and other public bodies, rather than an internal organizer for the settlers. This organization took place, to a greater extent, around milk collection and production.

3.2 "Born and raised in the countryside": the social characterization of family groups

The title of this item is quite thought-provoking, as it leads us to reflect on the characteristics of those beneficiaries who remain or those who, for different reasons, leave the rural settlements. In order to understand this behaviour, it is necessary to briefly characterize the family groups that benefit from the rural settlement policy in the Araguaia valley of Mato Grosso.

As has been said, understanding the composition of family groups, as well as their social characteristics, is an important variable in profiling the settlers who remain in the land reform areas. This section will therefore present variables relating to the age profile, gender, origin of the residents, among other issues related to interest in and form of access to the land. When analyzing the gender distribution in the settlement projects studied, it was found that there is not a great discrepancy between the two genders, to the extent that men represent 52% and women 48% in total.

However, when looking at each project, there is a certain disparity in this ratio, especially in the Volta Grande and Martins I settlements. The latter has a higher rate of women, while the Volta Grande project has a higher percentage of men. This gender disparity in the settlement becomes more evident when comparing it with the ratio in Brazil, in Mato Grosso and in the municipalities of each settlement. In the country, the percentage of women is 51%; in the state of Mato Grosso this percentage is 49%; in the municipalities studied, the percentage of women varies between 50% and 48%, i.e. in general there is a balance in the gender ratio, which is not perceived in the Martins I and Volta Grande settlements (IBGE, 2015).

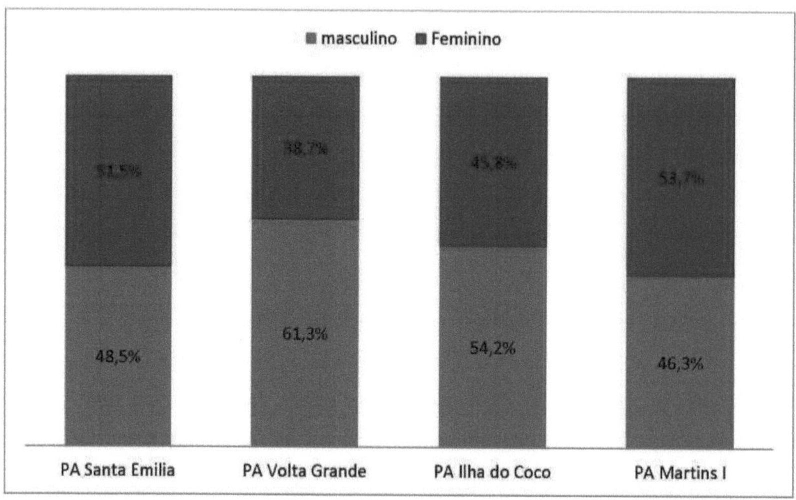

Figura 7. Gender profile of beneficiaries in the Settlement Projects, Araguaia Valley, MT, 2014.

Source: Prepared by the author, 2015.

Another important variable in understanding the social characteristics of the group is their age profile. The average age of the residents of the rural settlements in the Araguaia Valley of Mato Grosso is 43, with the highest age recorded being 84 and the lowest one year. An interesting way of interpreting the age profile is through stratification and, in view of this, it was decided to establish two categories of analysis, namely: individuals aged 0 to 14 and over 65, who make up the group characterized as the Economically Inactive Population (EIP); and the category aged 15 to 65, characterized as the Economically Active Population (EAP). According to the IBGE (2015), the concept of the Economically Active Population comprises the workforce available to the productive sector, i.e. the employed population, which is working, and the unemployed population, which does not have a job but is looking for work. The Economically Inactive Population includes people not classified as employed or unemployed.

It was found that, in general, in the four settlement projects studied, the economically active population represents 68% of the beneficiaries interviewed, while the EIP reaches 32% of cases. When analyzed individually, it is confirmed that in all the projects the EAP exceeds the 50% mark. It is worth noting that the Santa Emilia settlement has the highest rate of economically inactive population, with 47% of all beneficiaries interviewed, and the Martins I settlement has the highest percentage of economically active population.

Economically Active, with 77.5%. Figure 8 shows the age groups in detail in the four settlements analyzed.

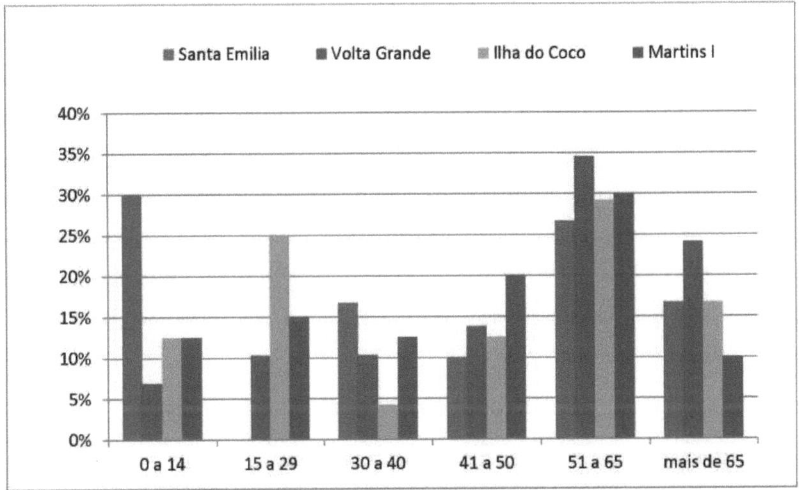

Figure 8. Age profile of the population living in the settlement projects, Vale do Araguaia, MT, 2014.

Source: Prepared by the author, 2015.

In the Santa Emilia, Volta Grande and Martins I settlements, the majority of family nuclei were made up of two people, representing 54%, 54% and 48% of the total number of families, respectively. In the Ilha do Coco PA, 40% of the family nuclei were made up of three people, and 30% were made up of two people. It is therefore clear that most of the houses are inhabited by the beneficiaries and their spouses, so that in the settlement projects, the family configuration of the permanents is generally made up of couples.

The data collected showed that, in general, the educational levels of the residents of the settlement projects were considerable, since 43% of those interviewed had completed elementary school, 21% said they were literate and 7% said they had no schooling at all. It was also observed that around 3.4% of the beneficiaries had completed higher education. Many of these cases represented municipal school teachers who were married to settled farmers. In addition, 10% of the population was studying and 2.5% of children were under school age. Therefore, roughly speaking, only 7% of those interviewed said they could neither read nor write.

Another variable that the data allowed us to assess is the origin of the beneficiaries. In this sense, it was found that most of them come from the state of Goiás. Also noteworthy is the large contingent from the state of São Paulo, which means that the rural settlements act as a mechanism for maintaining this migrant population that comes to the central region of the country in search of a better life.

Individuals whose place of origin was Mato Grosso itself, the state where the projects were based, had a much lower percentage, with no more than 18% of the total population. This may be justified if we consider the history of the occupation process of the Central West region, and especially of Mato Grosso, which was greatly influenced by government policies for occupying the national territory, through the use of land, which meant sending people to the "uninhabited" regions of the country. Thus, throughout the 1970s and 1980s, as explained in Chapter II of this work, there was a range of policies aimed at sending farmers to the central region, through the provision of credit for the development of agriculture. In addition, as a way of minimizing agrarian conflicts in the coastal regions, this conflicting population was relocated to the central region, so that these factors promoted a strong migration to Mato Grosso.

Table 1 shows the states of origin of the beneficiaries of the settlement projects studied.

Table 1. State of origin of the beneficiaries of the settlement projects, MT, 2014

State of origin	PA Santa Emilia (%)	PA Volta Grande (%)	PA Ilha do Coco (%)	PA Martins I (%)	Total (%)
Goiás	77	33	50	53	54
Mato Grosso	8	25	20	20	18
Minas Gerais	8	0	0	0	2
Sao Paulo	0	25	30	7	14
Rio Grande do Sul	0	0	0	13	4
Rio Grande do Norte	0	0	0	7	2
Bahia	8	8	0	0	4
Piaui	0	8	0	0	2

Source: Prepared by the author, 2015.

When analyzing the data presented in Table 1, it can be seen that the Santa Emilia settlement has the largest population from the state of Goiás. This may be due to the fact that the settlement is located in the municipality of Barra do Garças, which borders the state of Goiás.

of Goiás and is a transit route to both the Pantanal and the Western region and to the Eastern region of the country.

We also sought to understand the presence of relatives of the beneficiaries in other agrarian reform projects, or even within the settlement in which they live. As a result, 68% of those interviewed said they had a relative living in rural settlement projects, with the vast majority living in the respondent's own settlement. This is important due to the fact that most of the current beneficiaries acquired their plot through purchase, i.e. they were not selected by INCRA. As a result, the existence of relatives in the settlement projects can be considered an important attraction factor.

In the Santa Emilia project alone, the number of people selected by INCRA[20] is higher than those who have bought a plot, which is evidence of a weakened agrarian reform policy that has proved incapable of managing and supervising the projects and, above all, of curbing this parallel market for buying and selling land over the years. Figure 9 shows how the beneficiaries acquired their plots.

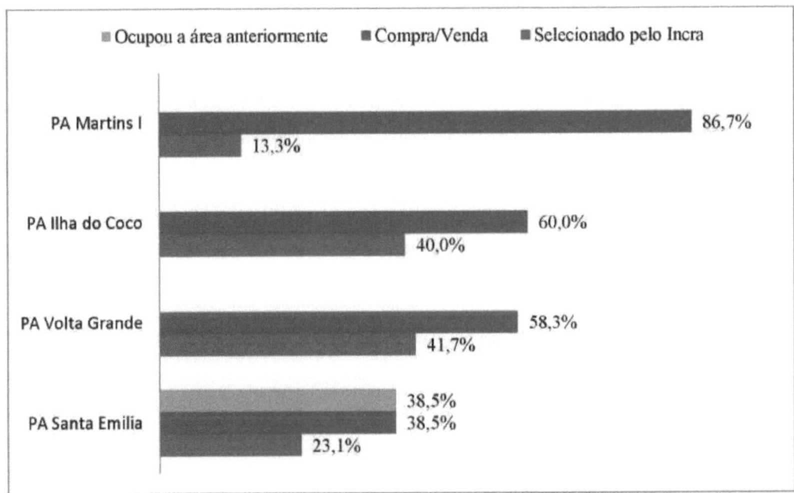

Figura 9. How residents of settlement projects in the Araguaia Valley, MT, acquired their plots, 2014.

Source: Prepared by the author, 2015.

According to Law No. 8.629 of 1993, which regulates agrarian reform, each beneficiary of the program will obtain a concession of real right of use - CDRU - for the plot, which is non-negotiable for a period of 10 years. In this way, the farmers covered by the program have the right to use the land, but are unable to sell this right during this period. As a result, the vast majority of parallel trade is illegal, and the federal agency is complicit, since there is an internal INCRA regulation that allows these people who bought plots illegally to be regulated if they meet the minimum

---

20 INCRA's selection includes both farmers who have previously occupied the plot and farmers who have registered to obtain the plot. In this sense, selection includes both mechanisms. However, the paper has been separated in order to present how these two models of inclusion are divided.

requirements for selection. In this way, the agency indirectly favors the offenders, since there is a register containing the names of people who could be allocated to settlement programs and are on the waiting list; however, these people are not allocated to available plots, because they were bought by others who, if they meet the requirements, are allowed to stay.

It's worth pointing out that in the older settlements, i.e. those with more than 25 years of existence, the majority of plots were acquired through purchase, showing that there is a tendency for the less adapted beneficiaries[21] , to sell the acquired plot as the years go by. As a result, only those who are better adapted to the environment remain living in these areas. Even if unintentionally, there is a "selection" that forces the settlers with the least conditions to remain, especially in terms of financial conditions.

As time goes by, the settlers who have no alternative means of survival are forced to dispose of their plots, leading to an increase in evasion in these areas. One way of analyzing this issue is to compare the length of residence with the time each project was created. In this case, the Santa Emilia, Volta Grande and Ilha do Coco settlement projects have a large number of residents who have lived within their boundaries since they were created. This situation was not found in the Martins I settlement, where there was a higher proportion of residents who said they only lived in the area after the project was created. Figure 10 shows the percentages relating to this question in detail.

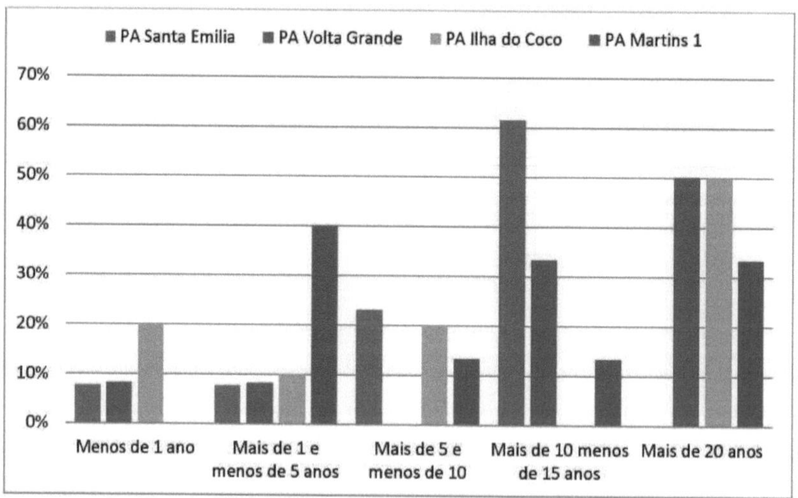

Figura 10. Beneficiaries' length of residence in the settlement projects, Vale do Araguaia, MT,

---

21 The word "adapted", in this context, is not limited to knowledge of the rural environment, but to factors that allow settlers to remain in the settlements. In this way, adaptation expands its meaning to encompass various factors that allow rural workers to stay in the countryside.

2014.

When analyzing the reason for these people coming to the settlement, a similar behavior can be seen in all four projects. Among the different reasons given, the beneficiary's desire to own land is the most influential factor, as exemplified by one interviewee:

Here we are our own bosses, there's no one to boss us around." (Resident of the Volta Grande Settlement Project).

The interviewees also pointed out that the limitations of not knowing what else to do, or having always lived in rural areas, are also reasons for these workers to seek a livelihood in these projects. This ideal, often generated within social movements organized by the struggle for land, which see the farmer as a revolutionary agent, is contradicted when it is observed that the search for land, by these settlers, occurs out of a desire for domination, not breaking with the logic of ownership; on the contrary, it reinforces it, insofar as they also want to own the property title. Figure 11 shows the different reasons that have influenced the arrival of workers in rural settlements.

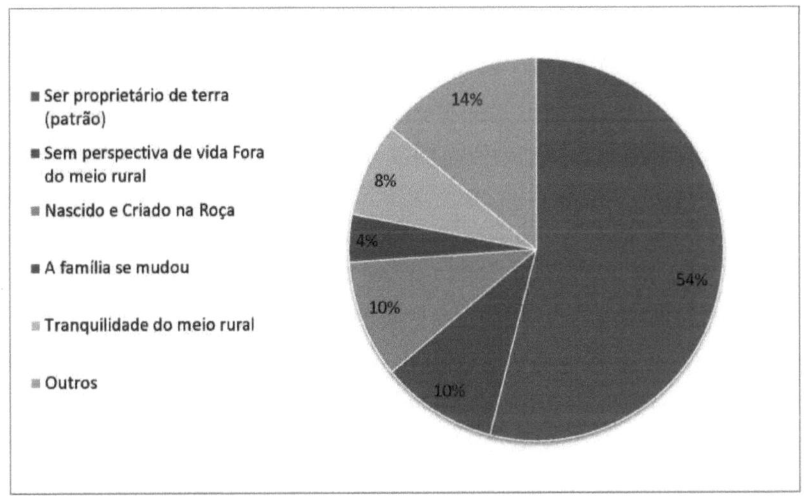

Figura 11. Reason for moving to the settlement projects, Araguaia Valley, MT, 2014.

The term "Born and raised in the countryside" accounted for only 10% of the total. However, even among those who said they wanted to own land, the justification of proximity to the countryside was still mentioned in the vast majority of cases. Experience of the countryside was often a central factor in the decision to move to the rural settlement projects.

3.3 "If you don't work outside, you eat the cattle". "Selection" in the agrarian reform projects of the Araguaia valley, economic conditions in rural settlements

This sub-item aims to highlight the economic characteristics of the beneficiaries of the settlement projects studied. This is a relevant issue, since the permanence of these settlers is linked to their economic conditions. Variables such as the main occupations of family members, source of income and other issues that make up this scenario will be presented. From this perspective, it is important to find out what type of occupation the settlers have, as well as their source of income and their experience with agriculture, in order to understand why they remain in the agrarian reform projects.

The main occupation of the workers in the four settlements surveyed is that of farmer, with 64% of those interviewed. This proportion is distributed differently between the projects, since in the case of the Santa Emilia settlement, 100% of the beneficiaries identify themselves as farmers, more significantly than in the Ilha do Coco settlement, where this category represents only 40% (Chart 3). In addition, professions linked to the urban sector, such as municipal school teachers, trade unionists and businesspeople were also mentioned, representing the diversity of trades that the settled community has, and breaking with the logic that rural settlements are only places for rural workers. It is also important to note that the category "retired" is significant in the Ilha do Coco settlement, and was identified by the interviewees as their main activity[22] .

Still looking at the main occupation of the settlers, it can be seen that in the older projects, such as Martins I and Ilha do Coco (both created in 1987), the diversity of occupations is greater than in the Santa Emilia settlement, which has only farmers as its main occupation. This once again demonstrates that, as time goes by, the need for these beneficiaries to look for new income opportunities increases, since the financial results from the agricultural activities carried out on the plots alone end up not being enough to maintain the family nuclei.

Thus, the precariousness of access to the production factor "capital" imposes a "selection" on those who depend solely on their plot to survive, which often results in the abandonment of agricultural activity and the consequent sale of the plot. It is clear from the results related to this issue that rural workers who are less well prepared or have fewer financial resources abandon/sell their plots and make room for workers who have better financial conditions, even if they are not connected to the rural environment. Table 3 below shows the details of this issue.

Table 3. Main occupation of the beneficiaries of the settlements studied, MT, 2014

Main occupation	Santa Emilia	Volta Grande	Coco Island	Martins I	Total (%)

---

22 This table only shows the main occupation of all the interviewees; there are other retirees who identified themselves as professionals in another field and are therefore not shown.

	(%)	(%)	(%)	(%)	
Farmer	100	63	40	48	64
Retired	0	0	20	8	6
Student	0	0	7	0	1
Professor	0	13	13	4	6
Entrepreneur	0	0	0	16	5
Snack	0	0	0	8	3
Nursing technician	0	0	0	4	1
Stonemason	0	13	0	4	4
Self-employed	0	6	0	4	3
Health Agent	0	0	0	4	1
Trade unionist	0	6	20	0	5

Source: Research data, 2014.

As well as identifying the main occupations of those interviewed, we also sought to find out where the beneficiaries worked. The data revealed that the main place of work is in the settlement itself, in most cases, considering the reality of the Santa Emilia, Volta Grande and Martins I projects. However, the Ilha do Coco project has a strong presence of professionals whose main place of work is in the municipality of Nova Xavantina, due to the presence of teachers and trade unionists living in this settlement.

This different dynamic, caused by the different professions, is an important factor in the beneficiaries' permanence in the settlements, considering the need to diversify their income, where dependence on the plot alone becomes unviable for their social reproduction. This becomes clearer when you realize that 16% of those interviewed had no experience whatsoever in rural areas at the time they came to the settlement, and yet they continue to live in the projects, largely because of the dynamic nature of their professions and sources of income. Figure 12 below shows the main places where the beneficiaries carry out their work activities, taking into account the four projects studied.

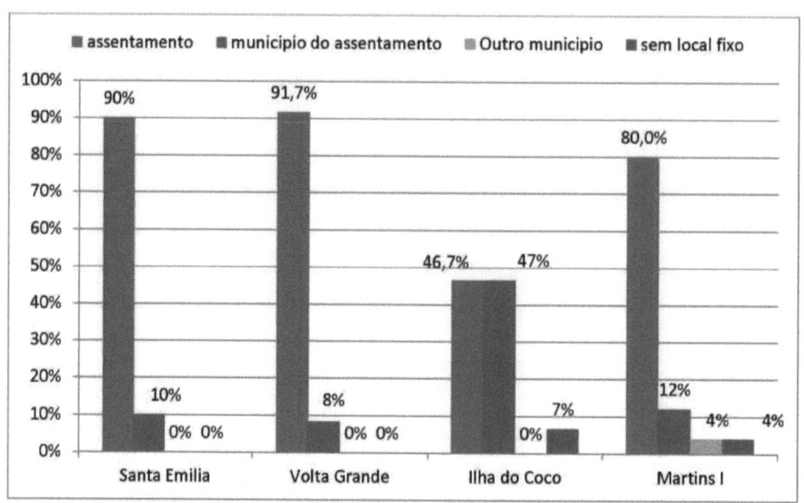

Figura 12. Location of the main occupation of the beneficiaries of the settlement projects, Vale do Araguaia, MT, 2014.

source: prepared by the author, 2015.

In the same way that the variety of professional occupations becomes a mechanism for persistence in the countryside, the source of income from activities carried out outside the plot is also an important factor in the permanence of these beneficiaries in the settlement projects. In this sense, 86% of the beneficiaries interviewed said they had an external source of income due to the need to diversify and supplement their income with different activities. Figure 13 shows the percentages of settlers who have external income and once again shows the disparity between the older settlements and the newer projects.

As has been said, the difficulty of consolidating productive activities, whether due to the unsuitability of the activity or the lack of "capital" to invest, forces the beneficiaries to diversify their sources of income. These factors are the key to understanding the permanence of the settlements in the Araguaia Valley of Mato Grosso. This need to diversify income is revealed in the fragility of the rural settlement policy currently in force in the country, which excludes farmers who have no other form of subsistence than what they produce on their property. As a result, the vast majority of the settlements surveyed are rural workers who need other activities to remain in the countryside.

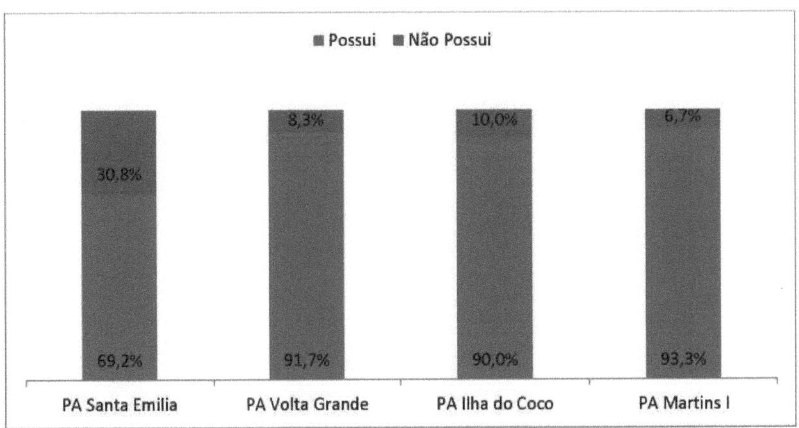

Figura 13. Existence of a source of income outside the plots for families in the settlement projects, Vale do Araguaia, MT, 2014.

Source: Prepared by the author, 2015.

This income, in the majority of cases, is related to activities such as retirement, in 53.5% of cases, providing services, in 19% of cases, and mainly in activities linked to the rural environment, such as cowboy, tractor driver and pasture cutter. Work as a civil servant, owner of a commercial establishment and owner of a rental property are activities practiced and represent 19%, 5% and 2%, respectively, of the total number of beneficiaries who said they earned an income outside the plot.

To get a better idea of this issue, we tried to identify separately the monthly income from activities carried out on the plot and activities characterized as external to the plot. Figure 14 shows the salary range from outside work. It can be seen that, in the three oldest settlements, the salary range is greater than in the Santa Emilia project. This is due to the greater occupational variation in these projects, which generates a diversified and higher income.

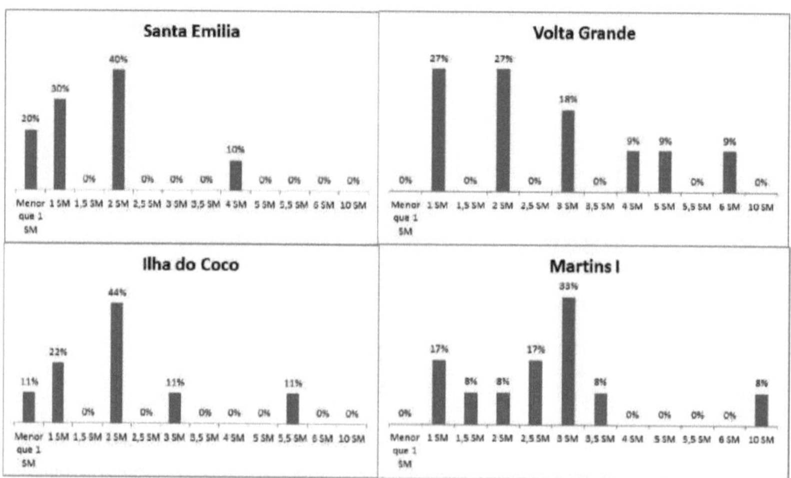

Figure 14. Value of income outside the plots in the settlement projects, Araguaia Valley, (in Minimum Wages - SM), MT, 2014.

Source: Prepared by the author, 2015.

It can be seen that in the Santa Emilia, Volta Grande and Ilha do Coco settlements, the highest percentage of external income is in the range of up to two minimum wages, at 90%, 54% and 77% respectively. Only in PA Martins I does the average external income increase, with 83% earning up to three minimum wages. The highest income found in the four settlements was in the Martins I settlement, where one settler claimed to earn around ten minimum wages, due to a commercial property in his home town. In general, the income outside the plot amounted to a maximum of three minimum wages, with occasional variations of more in some settlements.

It can be seen that a significant proportion of the settlers earn less than one minimum wage per month from the activities carried out on their plots, even exceeding 50% of those interviewed, as in the case of the Ilha do Coco and Martins I projects. In general, the average income from the activities carried out on the plots is in the range of up to one minimum wage for all the settlements. In the case of the Ilha do Coco settlement, one of the interviewees had an income of three times the minimum wage, showing the possibility of generating a significant amount of income on the plot.

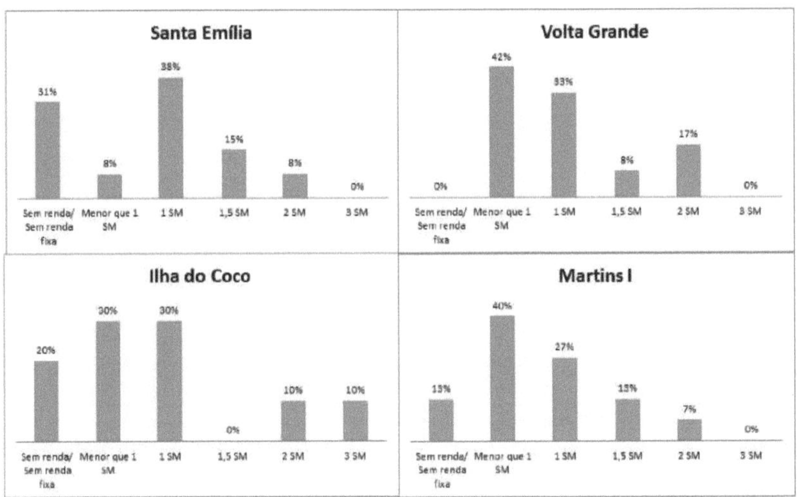

Figura 15. Value of income from plots in settlement projects, Vale do Araguaia, (in Minimum Wages - SM), MT, 2014.

Source: Prepared by the author, 2015.

However, when compared, income from outside the plot generates more resources and is therefore the main activity. In this way, it can be seen that in the settlements, permanent residents don't settle because they generate resources from agricultural activities, but because they diversify their source of income and look to multiple activities to guarantee their economic survival.

Thus, in general, the income earned from productive activities is of low value, although for the beneficiaries, measuring the production of subsistence items produced on the plot is imprecise. It can be seen that the activities carried out on the plots are secondary and act as a complement to income.

3.4 Cattle as savings and milk as a monthly salary: productive characterization of land reform beneficiaries

Production activities in the settlement projects can be divided into agricultural activities, related to plant production, and animal production, referring to livestock, pig, goat and chicken production. This distinction is also observed when comparing the use made of these activities. As far as crop production is concerned, the data shows that 84% of the settlers do it, 77% of whom produce only for their own consumption. Animal production, on the other hand, is practiced by 92% of the settlers and, of this total, 96% use their production for consumption and sale, showing a clear dichotomy of objectives within the settler's production chain: in the first case, production is aimed at subsistence and generating food; in the second case, as well as helping with subsistence, it generates economic income.

It is possible to see the variety of products grown in the settlements and their total percentage (Table 2). Although most of the settlers don't sell these products, they complement the family income, as the cost of food is reduced, since some of it is produced on the property. During the field visit, we observed a variety of foods that these families probably wouldn't have had with the same source of monthly income in the urban area; thus, even if they had the same financial conditions in the urban area, the plot provides a non-economic source of income, which can be seen in the availability of food, for example.

Table 2. Products grown in the settlement projects, Araguaia Valley, MT, 2014

Cultivated Products	Santa Emilia (%)	Volta Grande (%)	Coco Island (%)	Martins I (%)	Total (%)
Cassava	92	67	70	93	84
Cane	15	8	20	60	28
Banana	31	42	10	13	24
Corn	39	17	0	53	30
Horta	31	17	0	20	18
Gueroba	15	17	0	7	10
Coconut	0	8	10	0	4
Orchard	0	0	30	0	6
Courgette	8	0	10	7	6
Potato	0	0	0	7	2
Pineapple	23	0	0	7	8
Beans	8	8	0	0	4
Pequi	0	8	0	7	4
Lima	8	0	0	0	2
Rice	8	0	0	0	2
Mango	15	8	0	0	6
Cashew	15	0	0	0	4
Syringe	8	0	0	0	2
Cocoa	0	8	0	0	2
Orange	0	8	0	0	2
No plant	8	25	20	13	16

source: survey data, 2015.

Of the 16% of settlers who sell their agricultural produce, manioc flour is the only product sold in all the settlement projects. In the Santa Emilia PA, 8% of the beneficiaries sell manioc flour, as well as Gueroba (8%) and rubber extracted from rubber trees (8%). In the Volta Grande PA, 17% of the beneficiaries sell manioc flour, bananas (8%), corn (8%) and beans (8%). In the Ilha do Coco PA, 10% of the beneficiaries sold manioc flour, which was the only vegetable product sold by the interviewees. In the Martins I PA, 13% of the settlers sold manioc flour, along with rapadura (7%) and pequi (7%). In all cases, this trade is not regular, but is only practiced during agricultural production seasons. It is therefore an alternative way of supplementing income, although it is not a central income-generating activity for families.

This becomes clear when analyzing the area set aside for agricultural production: the largest area identified was nine hectares, in the Volta Grande settlement, while, in general, the average set aside for agriculture is two hectares in all the settlements. This confirms the cattle-raising nature of these projects, which allocate a large part of their land to cattle-raising activities.

The low level of agricultural activity is due to a series of limitations that the settlement projects have, whether it's the distance from commercial centers, which makes it difficult to transport production, or the low fertility of the soil, or the lack of access to credit for greater investment in these areas. Table 3 shows the main difficulties reported by beneficiaries in practicing agriculture in the settlements.

Table 3. Difficulties in agricultural production in the rural settlements surveyed, Vale do Araguaia, MT, 2014

Production difficulties	Santa Emilia (%)	Volta Grande (%)	Island of Coconut (%)	Martins I (%)	Total (%)
Low soil fertility	18	45	30	20	28
Lack of credit/financial support	9	9	10	33	17
Lack of agricultural machinery	9	0	10	20	11
Unfavorable weather / lack of water	18	0	10	7	9
No means of transport	0	0	10	0	2
Distance from shopping centers	27	0	10	7	11
Agricultural pests	9	18	0	0	6
No experience in agriculture (planting)	0	0	0	7	2

No difficulty	9	9	0	0	4
Lack of technical assistance	0	9	0	0	2
Not applicable	0	9	20	7	9

Source: Prepared by the author, 2015.

With regard to animal production, 92% of those interviewed said that they practiced this activity, distributed between beef and dairy cattle, chicken, pig, goat and fish production. Livestock is the central activity, practiced in all the settlements by 86% of the total, and also the one with the greatest impact on income generation. Chicken production is also very significant, with 80% of those interviewed practicing it in all the settlements, and it has become part of the food base for these families. The other existing livestock have little influence on the composition of the herd and can be seen in Table 4 in detail in each project analyzed.

Table 4. Animal production activities in the settlements surveyed, Vale do Araguaia, MT, 2014[23]

Animal Activity	Santa Emilia (%)	Volta Grande (%)	Coco Island (%)	Martins I (%)	Total (%)
Cattle	85	75	80	100	86
Chicken	77	75	70	93	80
Pork	54	25	50	67	50
Sheep	15	0	20	0	10
Fish	0	0	10	0	2

Source: Research data, 2015.

In the Santa Emilia and Martins I projects, dairy farming has a strong impact on the composition of income. This activity generates monthly income, unlike beef cattle farming, which offers sporadic financial gains throughout the year, due to the production of calves and the sale of breeding stock. Thus, in an analogical way, it can be said that beef cattle production functions as a "savings account", accessed only at certain times, while dairy production is more linked to obtaining continuous income. Dairy production is destined for dairies in the municipalities where the projects are located, which pay around R$0.65 per liter of milk every month, while beef cattle are mainly sold as old matrices and weaned calves, almost always at the weight of the arroba at the time[24] , to

---

[23] The total percentage for each settlement is more than 100% because it is possible for the same beneficiary to carry out more than one activity at the same time.

24 One arroba represents 15 kg. Cattle are generally sold by weight in arrobas and the price of this unit varies daily depending on international market demand. Its value is constantly changing.

occasional buyers or even neighbors.

On average, the area set aside for livestock farming is 26.5 hectares for the Santa Emilia project, 35 hectares for the Volta Grande project, 91.2 hectares for the Ilha do Coco project and 75 hectares for the Martins I project, revealing the importance of this practice, which occupies most of the total area of the plots. However, areas much larger than the official plot sizes were observed, thus revealing the acquisition of new areas by the beneficiaries[25] . The largest areas observed were 233 hectares, belonging to a beneficiary of the Ilha do Coco project, and an area of 225 hectares in the Martins I project.

Similarly, the production of poultry, mainly for free-range chickens and eggs, plays a significant role in generating income for the beneficiaries, since the surplus is usually sold sporadically in the region close to the settlement or in neighboring municipalities, without a fixed production scale. In this way, 24% of those interviewed said they sold their surplus production, earning prices in the range of R$20.00 to R$25.00 per chicken, and between R$4.00 and R$5.00 per dozen eggs.

Likewise, pigs are raised to subsidize the beneficiaries' own diet, and in some cases the surplus is sold. The average number of pigs kept per plot in the settlements is twenty-one in the Santa Emilia settlement, eight in the Volta Grande settlement, fifteen in the Ilha do Coco settlement and five in the Martins I settlement. The largest number of pigs was found in the Santa Emilia project, with one hundred animals on one plot. Most of them are sold by the kilogram of meat, which varies between R$4.00 and R$6.00 reais.

When it comes to goat breeding and fish farming, both activities are practiced, but by only a few beneficiaries and do not include the majority of the settlers surveyed. Sheep production was observed on a total of three plots, one in the Santa Emilia settlement and two in the Ilha do Coco settlement. In view of this, the rearing of this type of animal appeared to be aimed solely at family subsistence, without involving the sale of production. On the other hand, fish farming was found on one plot in the Ilha do Coco project, where a beneficiary had built a pond to produce fish, with a view to selling it. At the time of the research, sales had not yet begun, as the fish were still small, but the prospect was of weekly sales in the city of Nova Xavantina, at R$5.00 per kilogram.

The main difficulties experienced by these beneficiaries with regard to animal production are related to the lack of water in the Santa Emilia settlement, the low fertility of the soil in the Ilha do Coco and Martins I settlements and the lack of credit, according to the Volta Grande project settlers. In light of this, it is clear that the limitations to productive development, both agricultural

---

25 In some cases, the areas acquired by the beneficiaries were within the settlement itself, as in the case of a family that owned three plots, while in another case, the acquisition of new areas was made by buying areas neighboring the plots, which did not belong to the settlement. These situations were observed at PA Ilha do Coco and Martins I.

and animal, are strongly dependent on infrastructure conditions, such as distance from urban centers, low soil fertility, lack of water and lack of credit from the state, conditions that hinder agricultural production in the settlements.

Another important factor to note with regard to production activities is the labor used. The data collected revealed that in all four projects, family labor[26] is the most used; however, it is worth noting that in the Santa Emilia settlement, the hiring of day laborers played a significant role, as well as in the Ilha do Coco settlement. The interviewees said that they resorted to hiring day laborers or even exchanging days of service at specific times when the need for work increased, usually during harvest or cattle vaccination. In the Volta Grande settlement, there was even the case of one interviewee who had a salaried worker on a monthly basis to help with the work on the plot. The table below shows the percentages for the types of labor used to carry out farming activities.

Table 5: Labor used by beneficiaries in settlement projects, Vale do Araguaia, MT, 2014[27]

Labor	Santa Emilia (%)	Volta Grande (%)	Coco Island (%)	Martins I (%)
Family	92	83	80	100
Diarist	38	8	20	7
Monthly	0	8	0	0
Change of Day	8	0	0	0

Source: Research data, 2014.

In view of the above, it can be seen that livestock production activities stand out in terms of generating income on the plot, compared to, for example, plant production, which in most cases consists of meeting the family's subsistence needs. This productive logic, focused mainly on livestock, functions as a kind of savings account for the beneficiary, which is largely thanks to the source of income outside the plot, which ensures the basic monthly needs. In this way, cattle breeding serves as a supplement to the income of these settlers, and does not function as the family's main maintenance activity, which is left to the work or benefits, such as retirement, carried out outside the plot. This close relationship between external income and local production is one of the great characteristics that keeps this population in the rural settlements analyzed, since production on the plot is often insufficient to maintain the families' basic needs.

Similarly, dairy farming serves as a source of fixed monthly income for many beneficiaries who do

---

26 Family labor is treated as the activities carried out on the plot by members of the beneficiary's family, where there may or may not be payment for the activity.

27 The percentage in this case exceeds 100% due to the possibility of each interviewee answering more than one question about the labor used.

not have an outside income. In the Santa Emilia settlement, where there are fewer people with outside income, milk production generates monthly income for the farmers, so that the families' basic needs are met without the need for another outside activity. It is therefore the monthly income that largely guarantees that these families will remain in the countryside. This is clear from the words of one beneficiary about the need for a monthly income, as can be seen in the following excerpt:

Our expenses are every month, and the sale of calves alone can't pay my costs every month, I need to buy rice, beans, washing powder, pay for electricity? (Resident of the Santa Emilia settlement project).

Thus, what was observed is that if this monthly income is not met by agricultural production, or even by work carried out outside the plots, a recurring alternative is to rent areas within the plot to third parties, a practice known as "leasing", observed in all the projects. This occurs as a way of minimizing the need for resources and making use of areas that are underused. In the Santa Emilia settlement, 15% of those interviewed said they rented out an area equivalent to 31 hectares for the whole year - in other words, they rented out the whole plot. In the Volta Grande settlement, 33% said they rented an average area of 29 hectares for between seven months and a year. In the Ilha do Coco settlement, 10% of those interviewed rented four hectares for four months for farming. And in the Martins I settlement, 7% of the interviewees rented an area of 50 hectares for four months. Leasing has proved to be an alternative way of generating income for beneficiaries who don't exploit their entire plot, so the existence of an area with potential for leasing can guarantee families more annual income, which in turn can help them to remain in the settlement.

3.5 Infrastructure

Analyzing the infrastructure available to land reform settlers - both in collective terms and in the areas of the plots and those used for productive activities - is fundamental to understanding the conditions of the environment in which these workers find themselves. These aspects will be described in detail in this sub-item, looking at whether the compulsory credits of the agrarian reform programs have been released to the beneficiaries, or whether the collective infrastructure (such as roads, health posts and schools) has been built, or whether the conditions of the plot in terms of chemical and physical quality and housing conditions are at adequate levels to house a family.

According to the data made available by UAVA/INCRA, the credits under its responsibility were released in all the settlements and in 100% of the plots, with the exception of the Ilha do Coco project, which had only 44% of the plots receiving full credit and 50% of the plots receiving partial credit, totaling 94% of plots receiving credit (partial or full). INCRA is responsible for releasing funds for initial support, construction materials and recovery credit[28] . It was observed that in all the settlements analyzed there was a significant time lag between the families' access to the plots and the release of the funds, with a delay in the release of funds, creating a limiting factor for the development of the beneficiaries who often depend on these funds to start their agricultural activities.

In addition, access to private financing was practiced by 54% of the beneficiaries interviewed at the Santa Emilia project, 58% of those interviewed at the Volta Grande project, 50% of those interviewed at the Ilha do Coco project and 27% of those interviewed at the Martins I project. The vast majority of these loans are made through banks. The majority of these funds were used to buy cattle, both dairy and beef, in approximately 71% of cases. Improvements to the property's infrastructure, such as fences, corrals, dams and pastures, accounted for 14%. Other issues, such as health treatment for the beneficiaries, accounted for 10% of the funds spent, and in one case the funds were used for a 15th birthday party. In this sense, it can be seen that, of the number of settlers who accessed some kind of credit, the majority used the funds to improve conditions on the property or to increase productivity by buying breeding stock. It is therefore safe to say that there was a conscious use of the resources acquired.

Another important productive infrastructure factor that should be provided by the state is technical assistance. Interviews with UAVA technicians in Barra do Garças revealed that there is currently no technical assistance for rural settlements, and no other organization provides this type of service on an ongoing basis. According to Law No. 12.188 of 2010, which establishes the National Policy for Technical Assistance and Rural Extension for Family Farming and Agrarian Reform - PNATER, agrarian reform settlers are beneficiaries of this policy. However, what was observed in the research was the non-existence of this practice in the settlements, so much so that in the Santa Emilia and Martins I settlements, 100% of those interviewed said they had never received any kind of technical assistance. In the Volta Grande settlement, 17% of those interviewed said they had received a visit from an INCRA technician on their plot only a few times. In the Ilha do Coco settlement, 10% of those interviewed said they had received sporadic technical assistance from INCRA. When asked about the quality of the assistance provided by the agency, the settlers said it was of good quality, although not continuous.

Collective infrastructures such as schools, health centers and public telephones were only present in the Martins I settlement. This was due to the great distance between this project and its host

---

[28] For the three settlements in the 1990s (Volta Grande, Ilha do Coco and Martins I), the initial support was between R$355.00 and R$600.00 reais, the construction materials varied between R$629.00 and R$2500.00 reais and the recovery of construction materials was R$5000.00 reais,The latter was the same for the Santa Emília settlement, which differed only in the amount of the initial support, which was R$ 2400.00 reais, and the construction material, R$ 5000.00 reais.

municipality, Agua Boa, as well as the fact that it was located in a region of agricultural companies with a high number of resident families, such as the *Banco Safra* company, which even employed some settlers from this project. In this way, this structure helped not only the settlement itself, but the whole region. As already mentioned in the previous sub-items, this project had a school (up to the 3rd year of secondary education), a health center, a public ear and a community milk tank, all guaranteed by the municipal government, according to the interviewees themselves. All the other settlements had collective infrastructure in the host municipalities: the Volta Grande project in Araguaiana, the Ilha do Coco project in Nova Xavantina and the Santa Emilia project in Toricoeje, a district of Barra do Garças, located 15 km from the settlement. Although there were no structures of their own within the settlements, access was easy, especially for the children, who had school transport provided by the town halls.

In terms of housing, the settlements analyzed had good housing conditions. Table 6 shows the construction standard of the beneficiaries' houses, most of which are made up of masonry walls, clay tiles and cement floors. The vast majority have piped water and sanitary drainage is via cesspits. These good conditions are largely due to the housing loans provided by INCRA.

Table 6. Structures of beneficiaries' houses in rural settlement projects, Vale do Araguaia, MT, 2014

House structure		Total (%)
Walls	Masonry	88
	Madeira	12
Roof tiles	Clay	58
	Asbestos	42
Floor	Cement	64
	Ceramics	34
Water (piped)		92
Trench		96

Source: Research data, 2014.

A more detailed analysis showed that, in the Santa Emilia project, 92% of the houses were built less than 10 years ago. In the Volta Grande, Ilha do Coco and Martins I projects, the majority of the houses were built between 10 and 20 years ago, corresponding to 75%, 80% and 80% of those interviewed. These figures, when cross-referenced with the time when the settlements were created, indicate late construction, after the implementation of the projects. When asked about their satisfaction with the condition of their house, 50% of the interviewees were dissatisfied with their home, mostly because it needed finishing or more rooms.

With regard to the structure of the plot, the interviewees said that 70% of them considered the level of soil fertility to be good, 26% to be excellent and 4% to be very bad. Regarding the size of the plot, 70% considered it sufficient for the subsistence of one family.

Access to electricity was observed in 100% of the houses; 80% of the beneficiaries considered the roads to be reasonable; the presence of markets for buying and selling agricultural products was considered unsatisfactory by 62% of the interviewees; and the relief of the plot was considered satisfactory by 86% of the interviewees.

The beneficiaries' situation in terms of local infrastructure is generally satisfactory, with good quality housing, access to water and electricity in all the settlements. However, in terms of collective infrastructure, with the exception of the Martins I project, the others did not have it in the settlement, and it was necessary to move away from the settlement in order to use it.

15.4 Who leaves and who stays? Evasion in rural agrarian reform settlements

Evasion is a problem experienced in the settlement policy and is quite significant in the projects surveyed. According to the study *Percentages and causes of evasion in rural settlements* by Guanziroli et al. (2001), the state of Mato Grosso has an average evasion rate of 47.3%. Compared to the national average of 29.7%, there is a huge disparity in dropout rates in the settlements.

The researchers said that this evasion occurs because of the poor infrastructure available in the settlements, coupled with a strong absence of the state agencies that should be assisting these beneficiary families, as well as the great social differences between the settlers in terms of the possession of "capital", which in this case is not only material goods, but also cultural capital, such as study, experience in rural areas and kinship with other settlers. In this way, evasion cannot be understood as a single process, nor caused by a restricted range of factors, but as a set of situations that make it significant in agrarian reform projects (GUANZIROLI et.al., 2001).

In the four settlement projects studied, the average dropout rate was 57.3%, higher than the state and national averages. When asked if they knew of any residents of the settlement who had left, 96% of those interviewed said they knew of a case. In the view of those interviewed (in this case, the permanent ones), the main reasons for these people leaving were old age, which made it impossible to live in the countryside, poor health and the inability to receive treatment in the settlement, lack of financial resources, the illusion of living in the city and tiredness of staying in the settlement.

Therefore, for the permanent settlers, the lack of minimal structures and the lack of financial resources are the main factors related to the abandonment of plots in the land reform projects. Figure 16 below shows the main reasons for evasion in the opinion of the permanent settlers.

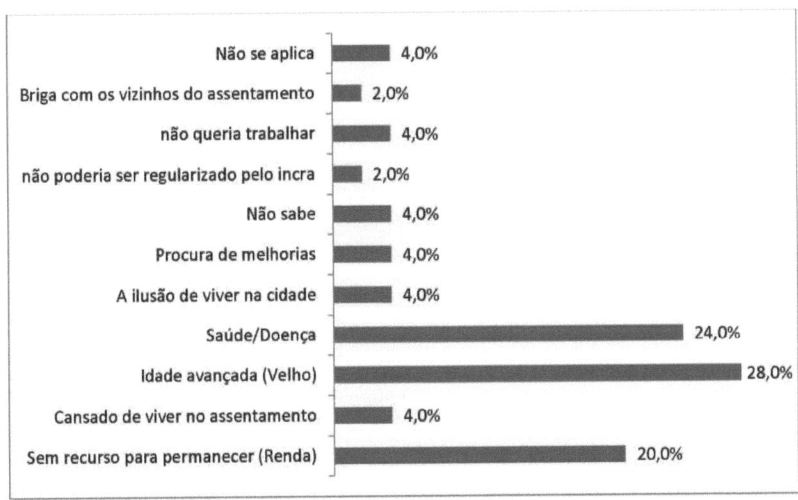

Figura 16. Reason for evasion of former beneficiaries of settlement projects, Vale do Araguaia, MT, 2014.

Source: Prepared by the author, 2015.

The interviewees were also asked to give their opinion on their desire to leave the settlement. In this case, 28% of those interviewed said they wanted to leave, with the Ilha do Coco and Martins I projects standing out from this total.

40% and 60% of those interviewed expressed this desire, respectively. The main reasons for this are homesickness (28%), difficulty in accessing healthcare (21%), and illness (14%).

What we see is a close relationship between the structural difficulties in the settlements and the precariousness of income generation on the plots, creating a situation of instability in which families without capital are more susceptible to abandoning the area. Difficulties in infrastructure alone, both collective and individual, are not the defining factors of evasion, nor is the absence of the state in terms of financial and technical support; what is noticeable is the existence of these conditioning factors together with financial fragility. And by financial fragility, we mean the absence of a fixed monthly source of income that guarantees the family's subsistence, which creates an environment conducive to abandoning the plot as an alternative.

To understand the mechanisms of permanence is also to understand the processes of evasion in rural settlements, and this work has shown that the existence of a monthly income that can cover living costs is a more important mechanism than the existence of minimum infrastructure or the presence of government agencies. This is because these difficulties are a reality throughout the Brazilian countryside and are therefore not determinants of the evasion process.

3.7 "My dream was to have land, now that I have it I won't let go". The profile of permanent beneficiaries in rural settlements

Tracing the profile of permanent beneficiaries in settlement projects is the main objective of this work. In this way, identifying the factors that allow these settlers to remain in rural areas is fundamental for the settlement policy, since keeping beneficiaries in the reformed areas is one of the major challenges of this policy.

Throughout this chapter, various characteristics of the beneficiaries were analyzed, from the composition and formation of the family group to the collective and individual infrastructures present in the settlements. Among all of these, there are some that significantly help workers to remain in rural areas, such as income diversification, which is nothing more than access to other sources of income than just the one that comes from the plot.

Access to an external income is one of the main characteristics of staying in the settlements, since it allows access to a monthly income, usually from work or retirement. Thus, as pointed out by Lamera & Figueiredo (2008), in a study of rural settlements in Mato Grosso, a large part of the income of the beneficiaries of this policy in the state came from complementary activities, such as day labor, harvesting, rural employment or handicrafts. This external income guarantees the family's economic stability, and agricultural production complements the income, so that compared to families with the same economic condition in the urban area, the settlers have a higher income.

Another factor for remaining in the settlements is the continuous activities that generate monetary income, i.e. agricultural activities that enable frequent income generation, such as milk production, which is supplied to dairies and produces a monthly income. In addition, the sale of chickens, eggs and manioc flour provides the beneficiaries with a constant source of income and helps the families' economic stability.

The question of infrastructure, in general, was not a determining factor in the decision to stay or leave, since settlements with better structural conditions, such as Martins I, had a higher dropout rate than settlements with worse conditions, such as Volta Grande and Santa Emilia. However, there was a greater frequency of young people in rural areas in the Martins I project, due to the fact that the school had a third grade of secondary education, which made it possible to keep these young people in the settlement. In this way, even if it is not a determining factor for staying, infrastructure creates the conditions for the family to remain and reproduce socially, indirectly helping in this process of keeping the beneficiaries in the rural settlements.

The interviewees pointed out the reasons that made them stay in the settlement: identification with the countryside, land ownership and tranquillity were the main reasons they gave. This shows that, in addition to the concept of land ownership on the part of these workers and their identification with the countryside, the notion of a better quality of life represented by the amenities to be found in rural areas, such as peace and quiet, was a factor in their staying. The idea that it is possible to reproduce oneself in a more qualified way on the plot is evident in the beneficiaries' comments. According to them, life in the countryside, even with its difficulties, offers a better quality of life than in urban areas. Quiet, as argued in many interviews, can be represented by the absence of the everyday problems of city life, such as crime, noise pollution and agitation. As such, the countryside provides an alternative to this context, thereby influencing the beneficiary families' decision to remain in the projects.

CHAPTER IV - EVALUATION OF THE VEGETATION COVER OF PASTURE IN RURAL AGRICULTURAL REFORM SETTLEMENTS

One of the issues pointed out in the analyses of evasion in rural land reform settlements, and also of the impacts of these projects on the environment, refers to the intensification of the process of deforestation and, consequently, degradation. In the case of evasion, the impossibility of carrying out deforestation is generally pointed out as one of the causes. Deforestation and the resulting charcoal production would guarantee settled families the income they need to get through the first few years and set up new production areas. However, as the vast majority of rural settlements are created in anthropized areas, mainly pastures in various stages of degradation, this possibility does not materialize. And in the absence of income and credit provided by the government, families flee.

Thus, in some cases, permanence can be attributed to the possibility of deforesting the areas of the family plots received. Despite the long-standing controversy over the inappropriateness of creating rural settlements to train new rural producers in sensitive or unsuitable areas for this productive activity, there is no precise information on the impacts of settlements on environmental degradation, especially in the Cerrado. In this sense, this part of the work will analyze the impacts of the settlements on the vegetation cover of the pastures in the areas where they are located. The aim will be to establish a correlation between the evolution of pasture quality, articulating economic and environmental variables, in the considerations of permanence in rural agrarian reform settlements.

This chapter will therefore look at the vegetation cover index of pastures, classifying them according to their state of degradation. This index provides information on the condition of the vegetation using satellite images. Through this technique, it is possible to see the vigor of the vegetation, its current vegetation condition, whether it has a high degree of degradation and water deficiency, for example. This type of analysis will be carried out using geographic information systems (GIS), such as Quantum Gis, and satellite images of the regions surveyed.

It will be possible to evaluate the land uses practiced by land reform beneficiaries over the years. It will also be possible to assess whether these rural settlements, due to changes in vegetation cover and the state of conservation of their pastures, can be linked to the process of environmental degradation in central Brazil.

The first part of the chapter will present the indices used, as well as all the tools and software used in this process. Subsequently, the entire methodological process for carrying out this work will be explained. Finally, land use will be discussed in relation to the permanence and degree of degradation of the pastures in the settlement projects.

4.1 indices and tools used in image analysis

Over the years, the use of spatial information tools and remote sensing has become part of the reality of the academic community and, in some cases, of society in general, as is the case with the use of GPS (Global Position System). Thus, greater access to this type of technology has made it possible for researchers in the most diverse fields of science to develop analysis techniques. The use of satellite images has made it possible to assess changes in the landscape over a given period of time, as well as classifying different land uses. In this study, the Normalized Difference Vegetation Index (NDVI) and Pasture Vegetation Cover (CVP) will be used as a tool to analyse the environmental evolution of rural settlements, which will be detailed below.

4.1.1 normalized difference vegetation index - NDVI

The vegetation index is a measure of the quantity and phenological condition of vegetation, made by analyzing spectral bands from remote sensors using digital image processing techniques. These indices analyze the spectral variation of vegetation and correlate it with its biophysical parameters. To do this, they generally use the red and near infrared bands, which account for more than 90% of the spectral variation in vegetation (SOUSA et al., 2007). Vegetation indices are an important tool for analyzing environmental changes in land use and have been used to estimate different vegetation parameters, such as leaf area index, amount of green biomass, evaluation of land use and management and recovery of degraded areas (LIMA et al., 2013).

Traditionally, vegetation monitoring using remote sensing has been carried out using vegetation indices. Among these, the Normalized Difference Vegetation Index (NDVI) has been one of the most widely used for estimating the amount of vegetation. This index is sensitive to the biophysical characteristics of vegetation and is fundamental for identifying changes in land use and cover (ANDRADE et al., 2011).

The Normalized Difference Vegetation Index (NDVI) makes it possible to map vegetation and measure its quantity and condition in a given area. NDVI is calculated from the red and near-infrared bands, which reflect electromagnetic energy from vegetation within these wavelengths. The physical principle on which NDVI is based lies in the spectral signature of plants: green, unrestricted plants absorb solar radiation in the red region as a source of energy for the photosynthesis process, while plant cells reflect strongly in the near infrared region. In this way, the difference between the reflectance of the red and infrared bands increases as a result of a number of factors, such as greener, better nourished, healthier plants without water restrictions, which absorb more red and reflect the infrared more intensely. Thus, the greener the vegetation, the greater the difference between the bands (INSTITUTO NACIONAL DO SEMIARIDO - INSA).

NDVI was proposed by Rouse et al. in 1974 and is defined by the following equation (MELO et al., 2011):

$$NVDI = \frac{IVP - V}{IVP + V}$$

Where:

NVI = Near Infrared Band, corresponding to reflectance at wavelengths between 0.76 and 0.90 micrometers.

V = Red Band, corresponding to reflectance at wavelengths between 0.63 and 0.69 micrometers.

Therefore, the results obtained by the NDVI equation vary between -1 and +1, with values closer to +1 representing denser and more vigorous vegetation, values close to zero representing less dense or bare areas, and negative values representing water or clouds, since the reflectance of water is greater in the visible range. The closer to the positive end, the greater the density of vegetation cover, and this value gradually decreases as density decreases, with positive but very low values. NDVI becomes important in this type of analysis, as it allows us to monitor seasonal and inter-annual changes in vegetation activity and development (ALVARENGA & MORAES, 2014; VIGANO et al., 2011).

4.1.2 Pasture Vegetation Cover - CVP

Livestock farming is an important activity in Brazilian agriculture. It generates thousands of direct and indirect jobs and plays a significant role in the country's gross domestic product (ANDRADE et al., 2013). In the rural settlements surveyed, this economic activity is found in all the projects and is practiced by 86% of the settlers interviewed, thus demonstrating its importance in the economic context of the beneficiaries. In view of this, analyzing the vegetation cover of pastures and their level of degradation is important for understanding the dynamics of these farmers and the way in which they exploit the land on which they live.

According to Gao et al. (2006) there is a linear relationship between the normalized difference vegetation index - NDVI and pasture vegetation cover - CVP, and this relationship is obtained by calculating the following equation:

$$CVP = \frac{(NDVI - NDVIs)}{(NDVIv - NDVIs)} \times 100\%$$

Where:

NDVI = Normalized Difference Vegetation Index;

NDVIs = the lowest NDVI value found among the pixels representing the area with exposed soil;

NDVIv = is the highest NDVI value found among the pasture pixels.

FMC values are classified into five degradation classes as shown in Table 7.

Table 7: Grassland vegetation cover classification values - CVP

1	Not Degraded	CVP > 90%
2	Slightly degraded	90 ≥ CVP > 75%
3	Moderately Degraded	75 ≥ CVP > 60%
4	Seriously Degraded	60 ≥ CVP > 30%
5	Extremely Degraded	CVP ≤ 30%

Source: Gao et. al., 2006.

Thus, as Gao et al. (2006) and Andrade et al. (2013) consider, it is possible to obtain the level of pasture degradation and establish its spatiality in relation to the total area surveyed. It is therefore an excellent tool for analyzing the degree of pasture degradation, as well as the evolution of this process over the years.

4.1.3 Atmospheric correction - DOS method

Satellites capture images at a great distance from the Earth's surface, and during this substantial journey, the electromagnetic radiation suffers different interferences from the atmospheric environment. In addition to this, there is the interaction between solar radiation and the radiation reflected by the target and the physical and chemical elements that make up the atmosphere, all of which is called Atmospheric Effects. These effects are represented by the phenomena of absorption and scattering. The former consists of the re-radiation of an electromagnetic wave by a particle present in the atmosphere, which redirects it to a different route from the original one. The phenomenon of absorption is based on the transformation of electromagnetic energy into another form of energy. As a result, the images obtained by satellite sensors lose sharpness and suffer surface brightness deformation, altering the measured signal according to the wavelength and composition of the atmosphere (NASCIMENTO, 2006).

In this sense, the atmospheric correction of satellite images becomes essential for image analysis, especially if a temporal analysis of the scenes is employed, since this methodology serves to equalize the image data on the same radiometric scale. This type of correction is used to minimize atmospheric effects caused by scattering, absorption and refraction of electromagnetic energy caused by water vapour, gases such as oxygen, ozone and carbon dioxide, as well as aerosols (small suspended particles) that are present in the atmosphere.

There are various methods for atmospheric correction that can be applied to satellite images, and they can be divided into alternative methods and physical methods. In the latter, it is important to obtain information on the optical properties of the atmosphere, as well as the process of radiation interaction with the atmosphere and the surface. Alternative methods, on the other hand, do not require atmospheric or surface data, but rather data from the image itself: they use the digital numbers that are present in the pixels and represent specific features, which can determine the interference of the atmosphere in the images (GOMES et al., n/d; SANCHES et al., 2011).

This work used the alternative method known as DOS (Dark Object Subtraction). This method was proposed by Chavez in 1988 and consists of estimating atmospheric interference using digital numbers (ND) from the satellite image, ignoring atmospheric absorption. In DOS, it is assumed that there is a high probability of dark pixels in the image, coming from the topography, or shadows caused by clouds, which should have a very low ND, equivalent to around 1% of the reflectance. However, these shaded pixels have higher ND values than expected, as a result of atmospheric scattering. They are therefore used as a reference for scattering correction: by analyzing the ND frequency histogram of an image band, a dark pixel value is chosen, then an atmospheric scattering model is selected, and the scattering value is estimated per spectral band. Subsequently, these scattering values are normalized according to the gains and *offsets* used by the imaging system to collect the data, so that the correction is made by subtracting the estimated values for the entire image (it is assumed that the scattering is uniform throughout the image scene, which is rarely true) (SANCHES et al., 2011). The DOS atmospheric correction is carried out according to the following equation (GOMES et al., n.d.):

$$\rho_{DOS} = j \times (ND - ND_{esp})$$

Where:

j = measurement estimated from the earth-sun distance, the solar elevation angle and the cosine of the solar zenith angle;

ND = digital numbers of the corrected image;

NDesp = atmospheric scattering calculated for the sensor.

In an article published by Silva (2013), in which the author compares a satellite image without atmospheric correction with the same image after it has been corrected using the DOS method, the influence that this type of image pre-processing has on the quality of the material used is clear. The IQI (Image Quality Index), an index that measures image quality, showed that there was a significant improvement in factors such as brightness, dark pixel adjustment and scattering correction. This method is therefore important in image processing for analysis.

4.2 Landsat satellites

The first Landsat (Land Remote Sensing Satellite) satellites appeared in the late 1960s as part of NASA's (National Aeronautics and Space Administration) land resources survey program. The first satellite in orbit was called ERTS-1 (Earth Resources Technology Satellite) and launched in 1972, and its successors were named Landsat. There are currently two satellites in operation, Landsat 7 (in poor condition) and Landsat 8 (INSTITUTO NACIONAL DE ESTADíSTICA Y GEOGRAFÌA - INEGI, 2014; ALVARENGA & MORAES, 2014).

These satellites are currently used by a wide range of actors, from scientists to the military, but the most frequent uses are for agricultural applications, forestry, land use analysis, hydrology, coastal resources and environmental monitoring, and are mainly linked to territorial studies in which the environment is the fundamental analysis parameter (COPPEL & LLORENTE, 2005). Since the first satellite in the ERTS-1 series (Landsat-1), a total of eight have been developed, of which only two remain active. Table 8 shows the chronology of the creation of the Landsat satellites and their time in use.

Table 8. Chronology of Landsat satellite launches

Satellites	Launch	Deactivation	Observations
Landsat 1	1972	1978	Called: (ERTS-1) and renamed to Landsat
Landsat 2	1975	1983	Called: ERTS-B and renamed to Landsat
Landsat 3	1978	1983	Known: Landsat - C
Landsat 4	1982	1993	Known: Landsat - D
Landsat 5	1984	2013	It was active for more than 27 years, three more than estimated.
Landsat 6	1993	1993	It never entered orbit
Landsat 7	1999	Active	First satellite to have a band pancromàtica (15 m)
Landsat 8	2013	Active	Use of Sensors: OLI and TIRS

Source: United States Geological Survey - USGS, 2014.

This study used images from the Landsat 5 and Landsat 8 satellites made available by the United States Geological Survey - USGS. The first, launched on March 1, 1984, has an equatorial orbit and an altitude of 705 km. It has a TM (Thermatic Mapper) sensor, which images the earth's surface, 185 km wide, with a spatial resolution of 30 meters and 7 spectral bands. The revisit time is 16 days for the same location, as can be seen in the

Table 9 (SILVA, 2002).

Table 9: Characteristics of the Thematic Mapper (TM) Sensor

Spectral bands	Band 1 - Blue (0.450 - 0.520 цт)
	Band 2 - Green (0.520 - 0.600 цт)
	Band 3 - Red (0.630 - 0.690 цт)
	Band 4 - Near Infrared (0.760 - 0.900 цт)
	Band 5 - Mid-infrared (1,550 - 1,750 цт)
	Band 6 - Thermal infrared (10.40 - 12.50 цт)
	Band 7 - Mid-infrared (2,080 - 2,350 цт)
Spatial resolution	Bands 1-5 - 30 meters
	Band 6 - 80 meters
	Band 7 - 30 meters
Imaged strip width	185 km
Time resolution	16 days

Source: SILVA, 2002.

Landsat 8, the last to be launched on February 11, 2013, has technological innovations compared to its predecessors. This satellite has an equatorial orbit and an altitude of 705 km, with a revisit time of 16 days (the same as Landsat 5 and 7). The sensors used are OLI (Operational Land Imager) and TIRS (Thermal Infrared Sensor) (USGS, 2014). These sensors, previously unseen in the Landsat series, have made advances such as the addition of band 1 (*new coastal),* designed to capture water resources and coastal regions, as well as a new infrared channel (band 9), for detecting cirrus clouds. In addition, the quantified radiometric resolution is 12 bits, as opposed to the previous 8-bit resolution, which improves the characterization of image targets and reduces shadow effects (KALAF et al., n.d.).

Table 10: Characteristics of OLI and TIRS sensors

Spectral bands	OLI
	Band 1 - (0.43 - 0.45 цт
	Band 2 - Blue (0.450 - 0.510 цт)
	Band 3 - Green (0.530 - 0.590 цт)
	Band 4 - Red (0.640 - 0.670 цт)
	Band 5 - Near Infrared (0.850 - 0.880 цт)
	Band 6 - SWIR 1 (1,570 - 1,650 цт)
	Band 7 - SWIR 2 (2.11 - 2.29 цт)
	Band 8 - Pancromatica (0.50 - 0.68 цт)
	Band 9 - Cirrus ((1.36 - 1.38 цт)
	TIRS
	Band 10 TIRS 1 (10.6 - 11.119 цт)
	Band 11 TIRS 2 (11.5 - 12.51 цт)
Spatial resolution	Bands 1-7 - 30 meters
	Band 8 - 15 meters
	Band 9 - 30 meters
	Band 10 - 11 - 100 meters
Imaged strip width	185 km
Time resolution	16 days

Source: United States Geological Survey - USGS, 2014.

The images from these two satellites differ in quality and technological innovation. However, for the calculations and operations we will carry out in this work, such as NDVI and CVP, we will only use the red and near infrared bands. In both satellites, these bands have a spectral resolution of 30 meters and it is therefore possible to compare the images generated by them.

4.3 Quantum Gis - QGIS

Remote sensing makes it possible to obtain spatial data such as images and digital maps that can be used in the planning, management, control and registration of the regions and zones analyzed. However, one of the major limitations of this type of tool is the Geographic Information System (GIS) software required to work with the data collected. In the vast majority of cases, the tools available are private and therefore require paid licenses to be used, which is often unfeasible due to the high cost of these products (OLIANI et al., 2012). As a result, the use of free software has become a viable alternative for those who want to work with this type of technology and do not have a large source of resources to do so.

At first glance, so-called Free Software (FS) may appear to be just free programs; however, they fulfill more functions than just the free nature of their product. The concept of FS is based on a series of factors regulated by the *Free* Software Foundation, which are: the freedom to run the program for any purpose; the freedom to study and improve the program, having access to the source code; the freedom to redistribute copies; as well as the freedom to improve the program and release it to third parties. As such, FS not only fit into the concept of free programs, but also allow users to alter and improve them themselves, which helps in the collective construction of information and knowledge (UCHOA & FERREIRA, 2014).

Quantum Gis - QGIS is free GIS software, written in C++ and Python, which makes it possible to visualize and work with geographic data through a digital interface. It has features for processing vector and matrix data, allows access to a library of GIS tools and also supports matrix formats such as ESRI, ERDAS and GEOTIFF. The QGIS project was started in February 2002 by the Open Source Geospatial Foundation (OSGeo), with the aim of generating a free viewer for geodatabases that would work on free operating systems (LINUX). However, over time, the program has become operative on all the main operating systems, supporting different vector and raster formats, providing an extensive database and geoprocessing functions (MANGHI et al., 2011; UCHOA & FERREIRA, 2014).

In short, QGIS is a robust geoprocessing tool that allows you to work with different file formats and has free access, which facilitates the work of small companies or public research institutions, which often have little budget to acquire this type of material.

4.4 Analysis methodology applied to evaluating pasture degradation in Settlement Projects

Spatial information tools and satellite images were used to analyze the current state of pasture in the settlements and compare it with the situation before the projects were created. The first part of this process was to obtain satellite images of the settlement projects. To do this, the USGS (United States Geological Survey) website was searched for Landsat images (Landsat 8 for 2014 and Landsat 5 for 2003 and 1990) for this analysis. The USGS is a scientific organization that makes archives of images from all over the world captured by satellites freely available. In this sense, the first step was to search for and refine images of the areas of interest, since many images have a high degree of cloud cover, which hinders and interferes with the classification of the proposed vegetation index. In addition, another important factor when it comes to classifying images and using indices is the time of year in which they are being analyzed. There is a strong relationship between the availability of water in the environment and the values obtained by the satellite, depending on the wavelengths. Therefore, in order not to overestimate or underestimate the results, it was decided to analyze three images of the same settlement at three different times of the year, an option that reduces the degree of error that may occur in the classification.

As each of the settlements analyzed in this dissertation has a different year of creation from the others, we looked for a date that could be common to all, so that it would be possible to represent the initial years of the settlement, with little interference from the beneficiaries in land use, in addition to obtaining good quality satellite images. Therefore, for the initial period for the Santa Emilia PA, images were taken from February 2001 and June and December 2003. For the Volta Grande PA, images were obtained from May 1988, January and June 1990. For the Ilha do Coco PA, images were taken in January and June 1990 and February 1992. Finally, for the Martins I PA, images were taken from January and June 1990 and April 1992[29] . In this way, a set of images was created at different times of the year, making it possible to generate a map that represented the normal or closest to normal conditions in each settlement during the initial period of its implantation.

Table 11: Period of the year in which the Landsat 5 images were captured in the region of the settlement projects, Vale do Araguaia, MT, 2014

Settlement	Initial images		
Santa Emilia	February 2001	June 2003	December 2003
Volta Grande	May 1988	June 1990	January 1990
Coconut Island	January 1990	June 1990	February 1992
Martins I	January 1990	June 1990	April 1992

Source: Prepared by the author, 2015.

Next, for the current images, 2014 was the most recent year in which it was possible to obtain images from three different times of the year, with good quality. Thus, the sets of images obtained for the four settlements surveyed, which represent their current situation, were from the months of February, July and December 2014.

After obtaining the images for the different periods of the year, and for the two comparative

---

29 There are images from different years in all the PAs. This is because, when images are obtained, the image provided by the USGS is often of poor quality and has a lot of cloud cover, which makes analysis impossible. In all the settlements, the best quality images were sought within a period close to a year.

periods, initial and current, the images were pre-processed in order to standardize them for comparative analysis, as well as reducing the reflectance effect that the atmosphere has on them. The method used for processing was Dark Object Subtraction 1 - DOSI. After standardizing the images, the vegetation indices were calculated, first the NDVI and then the CVP, on the images from each time of year. Three maps were therefore generated per settlement in the initial period and three maps per settlement for the current period.

Subsequently, a fourth map was generated from the sum of these three previous maps. In order to classify this fourth map and systematize it into the existing degradation categories, it was necessary to create an image interpretation methodology that would fit each pixel into a degradation class. This interpretative methodology consisted of setting pixel values for the maps of the different times of the year, so that the resulting map would be the sum of the values and each pixel value would be within a degradation category and could be identified. Table 12 below shows the pixel values set for each time of year:

Table 12: Pixel values per degradation category on the three different maps for each settlement throughout the year

Category	Map 1	Map 2	Map3
Extremely Degraded	1	3	5
Seriously Degraded	10	30	50
Moderately Degraded	100	300	500
Slightly degraded	1000	3000	5000
Not Degraded	10000	30000	50000

Source: Prepared by the author, 2015.

It can be seen that pixels with values referring to the sum of the same class on the three maps would have unique values and would therefore be easily classified. However, not all pixels throughout the year had the same degradation category, so pixel values referring to the sum of three different degradation classes were classified by the intermediate category. Pixels with values from two equal classes and a third were classified as belonging to the class of the first two. To better exemplify this classification system, Table 13 below shows the pixel values found in the fourth map and their respective classifications.

Table 13: Classification of the pixels in the fourth map according to the sum of the degradation classes in the first three maps[30] .

Pixel	Classes	Value	Classification
1	ED+ED+ED	9	ED
2	ED+SD+MD	531	SD
3	MD+MD+ND	50400	MD

Source: Prepared by the author, 2015.

At the end of this process, it was necessary to remove the "non-pasture" areas from the maps produced. This is because the CVP measures the level of degradation only in pasture areas; therefore, areas with woodlands and forests give rise to classification errors in the index generated by this calculation. To this end, using images from Google Earth (free software that allows you to visualize the globe on very small scales), areas with denser vegetation and forests were located in each project, so that they could be excluded from the analysis and quantification of the level of degradation in the CVP. It was assumed that the areas with forests and dense vegetation today would most likely also have been so in the initial years of the settlement's creation. In this way, selected "non-pasture" areas were used in both the current map and the map of the initial period[31] . Thus, at the end of all these procedures, the final map was drawn up, with the CVP of the settlements and their degradation categories.

---

[30] Degradation classes: ED = Extremely Degraded; SD = Seriously Degraded; MD = Moderately Degraded; ND = Not Degraded.

31 The division between two classification categories: *Pasture* and *Non-Pasture* areas, does not represent the entire system present in the settlements. This is because, in addition to these two categories, there were construction areas, watercourses, roads and dams that would not fit into this dualistic division. However, due to technical limitations such as the resolution of the image, the radiometric resolution of the pixel and the scale of the map, it was impossible to delimit and identify these areas. In view of this, it was assumed that both areas considered *Pasture* and areas considered *Non-Pasture* have components that do not belong to these classification categories.

Figure 17. Flowchart of the analysis and interpretation of satellite images of the rural settlements surveyed.

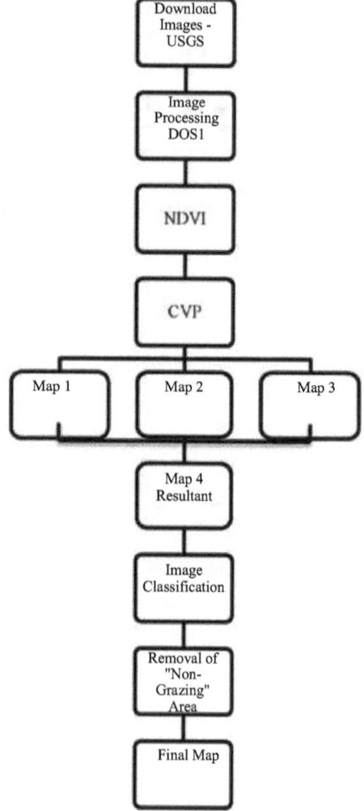

Source: Prepared by the author, 2015.

To better elucidate this methodological process, the flowchart presented in Figure 17 above shows the stages that were followed to carry out the image analysis and interpretation work:

4.5 The evolution of pasture in rural settlements. Reducing environmental degradation in land reform settlements in Mato Grosso

Understanding the factors responsible for farmers staying in rural areas, especially in the case of agrarian reform settlements, which are undergoing a process of precarious infrastructure, is fundamental to analyzing the country's settlement policy and its distortions in relation to the public who actually experience the reality of these projects. Among the possibilities for understanding these processes of staying in the countryside, land use over the years is an important mechanism for

82

evaluating the use of an essential element in any farmer's life: the land. One of the ways of analyzing land use can be through the level of pasture degradation, which is obtained by calculating the CVP on satellite images. Comparing the level of pasture degradation at two different times allows us to understand how the land has been occupied over time, and allows us to infer whether the practices employed by these beneficiaries are contributing to environmental recovery or to further degradation - the latter being very present and symptomatic in the Brazilian Cerrado.

The previous sub-item presented the methodological process for evaluating and analyzing satellite images of the degree of pasture degradation in the settlement projects surveyed. However, beyond just quantifying the level of pasture degradation, it is necessary to understand what it currently represents in the Brazilian Cerrado and its impact on the production process of Brazilian agriculture. Pasture degradation is one of the biggest problems facing Brazilian livestock today: of the approximately 60 million hectares cultivated in the central region of the country, it is estimated that 80% are in some stage of degradation. This problem directly affects the meat production chain, which in the rearing and fattening phase of cattle alone loses up to six times the production value of a well-maintained pasture or a recovered pasture (MACEDO et al., 2000).

The concept of pasture degradation can be defined as the evolutionary process of loss of vigor, productivity, the natural recovery capacity of pastures, the ability to overcome the effects of pests, diseases and invaders, leading to advanced degradation of natural resources as a result of inadequate management. The main causes of degradation are associated with: inappropriate type of forage variety for the site; inappropriate management and cultural practices, such as lack of soil preparation, lack of fertilization and use of fire; inappropriate animal management, such as overstocking; lack of soil conservation practices after prolonged use of pastures (KICHEL; MIRANDA & ZIMMER, 1999).

In view of this, the level of pasture degradation directly affects the population surveyed because, as we saw in the previous chapter, in all the settlements surveyed, livestock farming is an important activity, carried out by 90% of those interviewed, and in many cases it is the only source of income from the plot. This is because 66% of the farmers grow crops only for their own consumption, with animal production (mainly cattle, chickens and pigs) being the source of their agricultural income, as can be seen in Table 14 below:

Chart 14: Destination of agricultural production in the Settlement Projects surveyed, MT, 2014

Destination of agricultural production		
	Plant product (%)	Animal product (%)
Consumption only	66	4

Consumption and Sale	18	86
Does not produce	16	10

Source: Prepared by the author, 2015.

The results obtained by calculating pasture vegetation cover using satellite images in the settlements surveyed will be discussed below. The discussion will first be individualized, treating each settlement as its own. Subsequently, the common points between all the settlements analyzed will be analyzed, looking at the CVP and the level of pasture degradation over the years.

4.5.1 PA SANTA EMiLIA

The Santa Emilia settlement project was created in 2003 and is the youngest PA compared to the other three surveyed. In this sense, this site has its own particularities compared to the others, largely due to its short existence of around 12 years, such as the willingness of the beneficiaries to improve the site they are part of. The search for a source of monthly income was a central factor in the interviews and many pointed to the potential of dairy production to reduce the fragility of the monetary income obtained on the plots. In this sense, 85% of those interviewed in this settlement have livestock as a productive activity on the plot. And of this total, 45%, as well as working with beef cattle (selling the cattle), also work with milk production.

Livestock production is practiced by 92% of the beneficiaries. This shows the importance of farming for animal production, which in 77% of cases is destined for both consumption and sale, a fundamental factor in generating income. This becomes clear when we analyze the fact that 69% of the families' crop production is destined only for their own consumption, and therefore animal husbandry is the main source of income for the families in the Santa Emilia settlement. In fact, these activities demand a significant portion of the land available to the producers, especially when it comes to livestock. As a result, 67.3% of the total area of the settlement is made up of pasture, while the other 32.7% is made up of cerrado, forest and capoeira.

This is why it is important to analyze the degradation of vegetation cover in order to understand the current situation of the pasture, which is a fundamental part of the agricultural production process. More than just understanding the current situation, the temporal analysis makes it possible to compare the processes that have taken place in the environment, the development of degradation and its current stage, as well as generating a panorama between the beginning of the settlement and the present day.

Until 2003, the selected beneficiaries did not practice any type of farming on the property, even though they were within its boundaries after occupying the area. This allows us to state that, until this date, the environmental conditions of the farm were not the result of the activities of these farmers, but of the former owner and his predecessors. Field research showed that the property expropriated by INCRA was basically used for beef cattle, which is in line with the predominance of pastures when analyzing the vegetation cover in the settlement. In the initial period of the settlement, pasture conditions were 93.23% degraded, among its different classes. The most significant of these was *Seriously Degraded*, which accounted for 33.43% of the total pasture in the project, followed by *Slightly Degraded,* with 30.79%, *Moderately Degraded,* with 26.37%, and finally *Extremely Degraded,* which accounted for 2.64% of the total pasture area in the project. This data describes an area with high levels of degradation, containing a significant area in the classes with the highest degree of plant deterioration. Pastures considered *Not Degraded* occupied only 6.77% of the pasture area in the settlement during this period.

Over the years, there has been a process of recovery of pasture degradation in the Santa Emilia settlement. This is evident when analyzing the most degraded areas in 2014, such as the *Extremely Degraded* and *Seriously Degraded* classes, which now occupy 0.46% and 21.14% of the pasture surface, respectively. The *Moderately Degraded class now occupies* 35.79% of the pasture, while the *Slightly Degraded class* represents 36.57% of the total pasture area. However, this process of improvement in the pasture's degradation conditions was not reflected in the pasture classified as *Non-Degraded,* which fell by around six hectares to occupy an area of 6.04%. In

Table 4 shows the percentage of area occupied by each class of degradation in relation to the total pasture in the initial period of the settlement and today.

Table 4: Classes of pasture degradation in PA Santa Emilia, in Percentage, in the initial period (2001 - 2003) and currently (2014)

Phase	Initial (2001 - 2003)	Current (2015)
Class	Area (%)	Area (%)
Extremely Degraded	2,64	0,46
Seriously Degraded	33,43	21,14
Moderately Degraded	26,37	35,79
Slightly degraded	30,79	36,57
Not Degraded	6,77	6,04
Total	100	100

Source: Prepared by the author, 2015.

This process of pasture recovery during the years of occupation by the beneficiaries shows that the cultivation practices applied had a less harmful impact on the environment than those practiced by the previous owners. Figure 18 shows the vegetation cover of the pasture in the settlement in the initial period of its creation and the vegetation cover today. You can see a decrease in the degradation index in the eastern part of the settlement, where the old farmhouse was located, which was probably the place with the most livestock activity in the past. The map of the evolution of plant biomass over the period of the settlement's existence shows where there has been an increase in plant biomass in the pasture and where there has been a loss of vegetation: it can be seen that the increase in cover has occurred more uniformly throughout the settlement, while the loss has occurred more intensely in the southern part of the project, showing greater specificity in this location. It can be inferred that in this region, agricultural practices have had a greater impact on the environment, or even that the activities have been more intense, causing this environmental deterioration.

Figure 18: Pasture Vegetation Cover in the Initial Period (2001 - 2003) and Current Period (2014) and Map of Vegetation Biomass Evolution in the Santa Emilia Settlement, MT

Source: Prepared by the author, 2015.

4.5.2 PA VOLTA GRANDE

The process of creating the Volta Grande PA began in 1988, when the beneficiaries were allocated their respective plots, although the formal creation process was only made official in 1991. Thus, the use of this property by the families formally settled in 1991 had been going on since the late 1980s. This project has thirty-five families settled on plots ranging from 21 to 82 hectares. The productive activity that generates income for the vast majority of families is beef cattle, with 75% of those interviewed. Agricultural production is carried out in 75% of cases, but only 22% of families sell their surplus. Thus, livestock farming is the main income-generating activity for the settled families. Given this situation, the vegetative state of the pastures becomes an important factor in the continuity and economic viability of the activity in rural areas, since it is a central element in the livestock production chain.

In 1990, the settlement was in its third year of implementation; however, as the creation decree was only issued in 1991, and the credit was released in 1995, it can be considered that the beneficiaries' actions in terms of transforming the space and land use in this initial period were less intense, which allows us to analyze the situation of pasture vegetation cover as being very close to that of the period before the settlement was created. In this period, the classes of pasture degradation with the greatest surface area were *Slightly Degraded, with* 50.10% of the area, and *Moderately Degraded,* with 45.88%, thus representing 96% of the total pasture area in the settlement. The most degraded classes, such as *Seriously Degraded* and *Extremely Degraded,* together accounted for 0.45%, so that 96.5% of the pasture area was under some level of degradation. Pasture characterized as *Not Degraded* occupied an area of 3.50% of the pasture area in the settlement.

The classification considered as *Non-pasture*, which involves dense cerrado, forests and capoeiras, represented 18.42% of the total surface area of the settlement, so that pasture covered 81.58% of the surface area of the Volta Grande settlement. During the period, there was an increase in the most degraded classes such as *Extremely Degraded* and *Seriously Degraded,* which now occupy 0.13% and 7.87% respectively, an increase of 7.5% compared to the initial phase. The *Moderately Degraded* class increased to 46.45%, remaining practically stable, and the *Slightly Degraded* class decreased by around 10% to occupy 40.51% of the pasture surface. It is therefore clear that there has been an increase in the level of degradation of the pastures, which were already suffering from some form of plant deterioration: anthropogenic action has accelerated the process of wear and loss of existing pasture biomass. However, another important factor to note was the increase in *Non-Degraded* pasture, which now occupies 5.04% of the pasture area in the settlement. In view of this, it can be said that there were two distinct processes in the Volta Grande PA: the first, intensification of degradation in pastures with some stage of degeneration, especially the more intense categories; and the second, an increase in the area of undegraded pasture with greater plant biomass.

Table 5: Percentage of pasture degradation in the Volta Grande PA in the initial period (1988 - 1990) and today (2014).

Phase	Initial	Current
Category	Area (%)	Area (%)
Extremely Degraded	0,01	0,13
Seriously Degraded	0,44	7,87
Moderately Degraded	45,88	46,45
Slightly degraded	50,10	40,51
Not Degraded	3,57	5,04
Total	100	100

Source: Prepared by the author, 2015.

This movement in the extremes - both the intensification of degradation and its reduction - can be understood very much as a function of the activities carried out by the beneficiaries in the settlement. As explained in Chapter III, one of the major problems with the current settlement policy is the lack of investment and availability of credit for the settlers, and this difficulty is no different for the Volta Grande project, which has 42% of the interviewees without access to credit or financing throughout their stay.

This reinforces the idea that the practices developed by farmers on their plots are carried out with little or no resources and without the use of agricultural inputs and machinery. In this way, decapitalized farmers are unable to maximize their production and occupy their entire land, settling only in certain areas of the plot, intensifying their use and eventually increasing the process of vegetation degradation, while in other areas of the property, with little or no anthropogenic action, the recovery of vegetation cover is taking place. Areas with formed pastures are used more than others, which ends up leading to greater intensification of use, making regeneration more difficult.

Figure 19 shows the vegetation cover maps of the pasture in the initial phase and in the current phase of the settlement, as well as the map of the difference between the phases, showing the places where there was an accumulation of plant biomass and where there was a loss of vegetation in this period between the creation of the settlement and 2014. It can thus be seen that the loss of plant biomass from the pasture occurred in greater quantity and in a more generalized way, with greater intensity in the northern part of the settlement, while the gain in vegetation occurred mainly in the regions close to the Araguaia River, in the western part and in the eastern part of the project. When quantifying the changes that have taken place, it can be seen that in 31% of the pasture area there has been a loss of plant biomass, and consequently an increase in the degenerative process, while in 20% of the area there has been an increase in plant biomass and a decrease in degradation. The remaining 49% of the pasture remained unchanged throughout the study period.

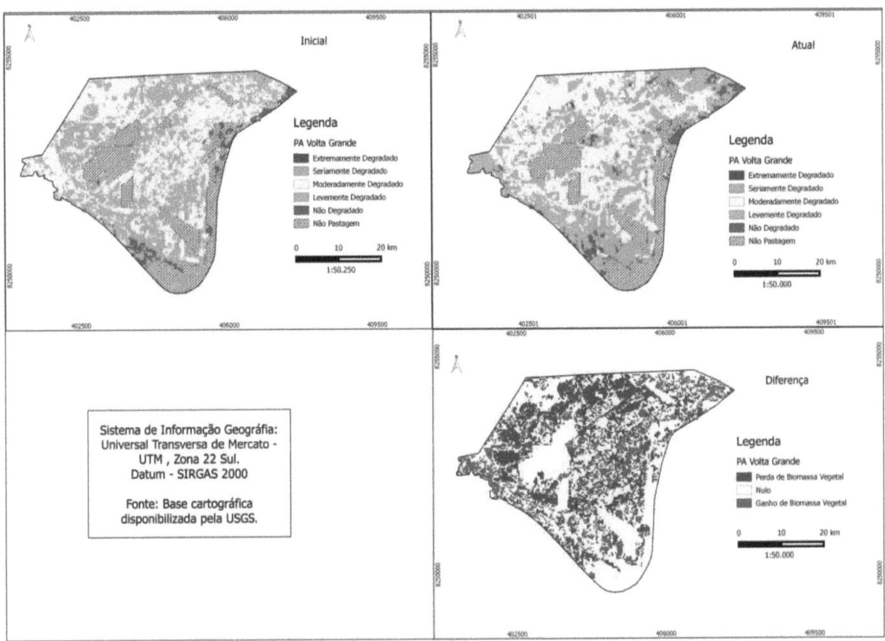

Figure 19: Pasture Vegetation Cover in the Initial Period (1988 - 1990) and Current Period (2014) and Map of Vegetation Biomass Evolution in the Volta Grande Settlement, MT

Source: Prepared by the author, 2015.

4.5.3 PA COCONUT ISLAND

The Ilha do Coco settlement project was created in 1987, regularizing an area previously occupied by farmers. There are 35 plots, ranging in size from 25 to 98 hectares. Cultivation is an activity carried out by 80% of the settlers, and the vast majority of production is destined for subsistence, comprising 87.5% of cases. However, animal husbandry is the agricultural activity that generates income: 90% of the families produce animals, 100% of which are for consumption or sale. Of this amount, 89% of the families are engaged in livestock farming, making it the activity with the greatest presence in the settlement and the main source of income. Other activities such as raising chickens, pigs, sheep and fish are present in 78%, 56%, 22% and 11% of cases, respectively, and are also sources of income because, according to all those interviewed, the surplus production is always sold.

It is therefore animal husbandry activities that have the greatest impact on the settlers' family economy. As such, the condition of the vegetation is of significant importance in the context of this

89

settlement. Pasture degradation influences the productive development of this community and is a limiting factor for economic expansion. The reality of the Ilha do Coco PA in the initial phase had a total degradation index of 92.72%, with the *Slightly Degraded* class having the largest area, with 51.86%, followed by the *Moderately Degraded class,* with 35.12% of degradation. The *Extremely Degraded* class had a negligible area and represented 0% degradation, while the *Seriously Degraded class* occupied 5.73% of the pasture surface. The pastures categorized as *Not Degraded* occupied an area of 7.28% and, in addition, the areas classified as *Not Pasture* comprised 9.66% of the total perimeter of the settlement, explaining the cattle-raising nature of this property, which had more than 90% of the total area in the form of pasture.

Over the course of the more than 24 years of existence of the Ilha do Coco settlement, the pasture vegetation cover has seen a significant recovery in its degradation, improving its indices in all classes. The area of *Non-Degraded* pasture jumped to 16.8%, an increase of another 9%, which in terms of area means an increase of 242 hectares. In addition, there were improvements in all the degradation classes: *Seriously Degraded* went up to 2.54%; *Moderately Degraded* went up to 24.66%; and *Slightly Degraded* went up to 55.99%, highlighting this process of pasture recovery over the years. Thus, the most intensely degraded areas decreased, while the mildly degraded areas increased.

Table 6: Category of pasture degradation in the Ilha do Coco PA in Percentage in the initial period (1990 - 1992) and currently (2014).

Phase	Initial	Current
Category	**Area (%)**	**Area (%)**
Extremely Degraded	0,00	0,00
Seriously Degraded	5,73	2,54
Moderately Degraded	35,12	24,66
Slightly degraded	51,86	56,00
Not Degraded	7,29	16,80
Total	100	100

Source: Prepared by the author, 2015.

It should be noted that this reduction in degradation can also be seen when comparing the gain in plant biomass of the pasture, which was 75% compared to the loss, which was only 25% between the initial phase and the current phase. This amount represents 50% of all the pasture in the settlement that has undergone some kind of change, while the other 50% has remained stable in terms of its category and level of degradation. Figure 20 shows this improvement in plant biomass and the decrease, especially in the most degraded categories.

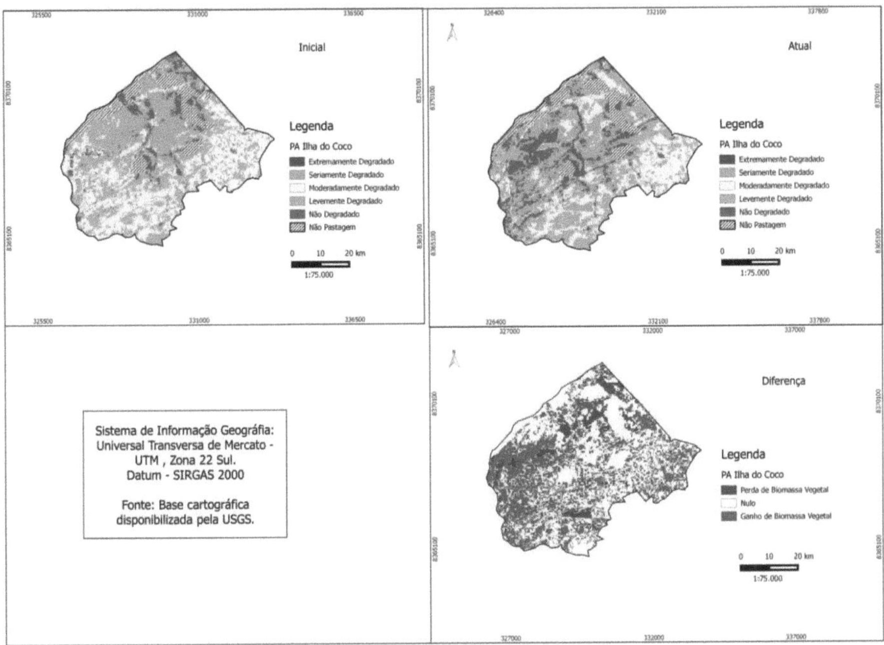

Figure 20: Pasture Vegetation Cover in the Initial Period (1990 - 1992) and Current Period (2014) and Map of Vegetation Biomass Evolution in the Ilha do Coco Settlement, MT

Source: Prepared by the author, 2015.

4.5.4 PA MARTINS I

The Martins I settlement came into being at the end of the 1980s, in 1987, but was only officially created in 1995. The settlement has fifty-five plots, ranging from 61 to 78 hectares. Just like the other settlements surveyed, animal production is the main productive activity practiced by the beneficiaries, and is their main source of income.

Animal husbandry is practiced by 100% of the settlers interviewed, and is intended for consumption and the sale of surpluses by the families. The main activity is beef farming, practiced by 100% of the families, followed by chicken production (93%) and pig farming (67%). Another factor that reinforces the livestock-oriented nature of the settlement is crop cultivation, which is limited to subsistence production in 77% of cases, with no profit motive. In this way, most of the income-generating activity comes from animal production.

In short, PA Martins I has similar production characteristics to the other settlements, in which the need for a good quality pasture area influences the activities of these families. Therefore, the area composed of *NON-pasture* represented only 13.76% of the total area of the settlement, so that 86.24% of the surface was pasture.

In the initial period, the characteristics of the settlement were as follows: 49.27% *Moderately Degraded* pasture, 31.21% *Seriously Degraded* pasture, 15.35% *Slightly* Degraded and 2.73% *Extremely* Degraded.

Degraded, resulting in a total of 98.56% of the pasture area having some kind of degradation. This is a high percentage, even when compared to previous settlements, and reveals the precarious state of the vegetation in this settlement at the time it was created. This fact reveals the lack of concern on the part of the competent bodies for settling landless farmers in areas with the quality and conditions for their socio-economic development.

In the current phase, the settlement has seen an improvement in the vegetation condition of its pastures. This is because the area of *Non-Degraded* pasture increased to 3.25%, and *Slightly Degraded* pasture now occupies 22.82% of the area, showing an improvement in the conditions of degradation. However, when analyzing the classes with the greatest impact, such as *Extremely Degraded* and *Seriously Degraded*, they remained unchanged in terms of area, occupying 2.66% and 31.67%, respectively. This shows that the areas where there had been moderate or slight degradation have recovered over time, a fact that can be seen in the reduction of the *Moderately Degraded* class by almost 10%, now occupying 39.60% of the pasture area.

Table 7: Category of pasture degradation in PA Martins I in Percentage from the initial period (1990 - 1992) to the present (2014).

Phase	Initial	Current
Category	**Area (%)**	**Area (%)**
Extremely Degraded	2,73	2,66
Seriously Degraded	31,21	31,67
Moderately Degraded	49,27	39,60
Slightly degraded	15,35	22,82
Not Degraded	1,44	3,25
Total	100	100

Source: Prepared by the author, 2015.

In this way, the creation of the settlement made it possible to recover areas with a lower intensity of degradation, due to the low level of exploitation and less intensive use of the soil by the farmers. However, the most degraded areas have remained stable, since in these regions specific recovery practices are needed, as well as the use of soil correction and fertilization, which is not practiced by the settlers, given the limited availability of resources and capital. Figure 21 shows the maps of the initial and current vegetation cover, as well as the map of the evolution of plant biomass over this period of time. It can be seen that there has been an increase in plant biomass in 33% of the pasture area, while 24% of the pasture has experienced plant loss. In addition, 43% of the pasture remained stable, with no change in the level of degradation.

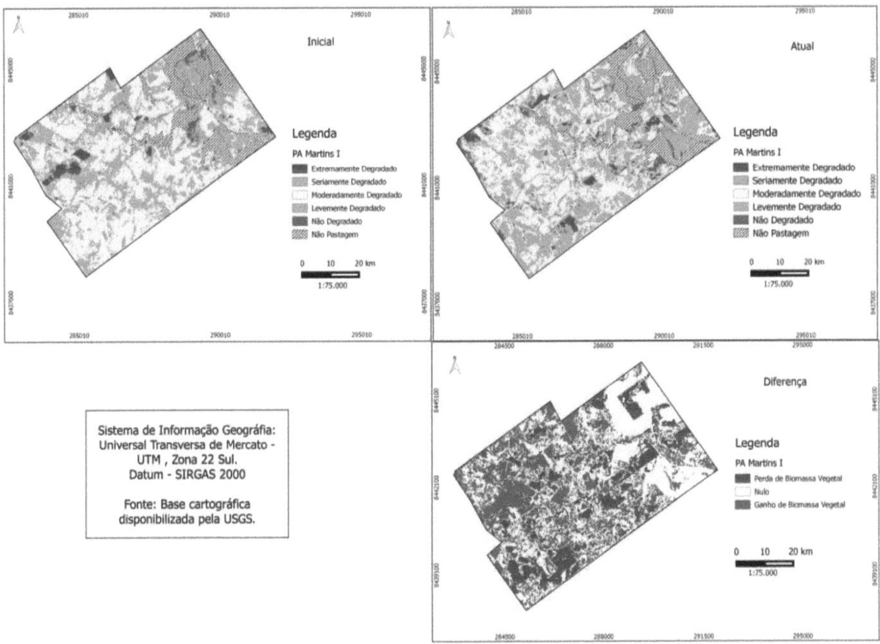

Figure 21: Pasture Vegetation Cover in the Initial and Current Period (1990 - 1992) and Map (2014) of the evolution of vegetation biomass in the Martins I settlement, MT

Source: Prepared by the author, 2015.

4.6 Recovering degraded pasture or increasing plant biomass?

In view of the above, the rural settlements surveyed generally increased their plant biomass index, as well as reducing pasture degradation. In all the projects, there was an increase in the *Non-Degraded* class, with the exception of the Santa Emilia settlement, which remained stable in terms of the area of pasture in this class. However, it is important to note that this improvement did not occur uniformly in all the settlements. As mentioned above, there were projects that improved the

93

levels of degradation in all categories evenly, such as the Ilha do Coco project; however, in other settlements, at the same time as there was an increase in non-degraded areas, there was also an increase in areas with more intense classes of degradation, such as the Volta Grande project. In addition, the Martins I settlement remained with the area with the greatest stable degradation, while the area of undegraded pasture increased.

In view of this, an analysis of the impacts of agrarian reform settlements on the environment cannot be carried out in a purely quantitative way. The figures collected show a partial improvement in the plant biomass of the pastures in these rural settlements, located in the Cerrado of Mato Grosso; however, the specific characteristics of each settlement and its internal dynamics, together with the type of farming carried out by the families, have a direct influence on the degenerative process of the pasture. According to Macedo et al. (2000), when pastures are in a process of degradation, they need to be renovated or recovered, the latter of which can be done directly through agronomic, chemical or mechanical practices. These practices improve the pasture by adjusting the animal stocking rate to the type of soil and adapting the management system to the desired productivity, as well as applying topsoil correction. In this way, it is possible to recover the pasture without destroying the vegetation.

In the case of the settlements surveyed, the major factor in pasture recovery is basically the less intensive practices employed by the settlers. When comparing the number of Animal Units (AU)[32] per hectare in the settlement in relation to the regional average, it becomes clear that the management practiced by the settlers is more environmentally sustainable[33] . This is because, according to the Mato Grosso Institute of Agricultural Economics (Imea), in the Barra do Garças micro-region, which also includes the municipality of Araguaiana, the average number of animal units per hectare is 0.60 AU/ha. In the Agua Boa micro-region, which also includes the municipality of Nova Xavantina, the average animal unit per hectare is 0.52 AU/ha. When we looked at the average number of animal units per hectare in the projects, we found that this index was equal to or higher than the regional average in all the settlements, except for the Ilha do Coco project. Table 14 shows the average plot area, average herd size and animal unit per hectare for each of the settlements analyzed.

Table 15 - Average plot area, average cattle herd, average animal unit per hectare in the settlement projects, Vale do Araguaia, MT, 2014

Average	PA Santa Emilia	Volta Grande PA	PA Ilha do	PA Martins I

32Zootechnical measure that characterizes the capacity of pastures to support animals.
33 One AU is equivalent to 450 kg of live weight. It was assumed that all the cattle in the settlement had a live weight of 450 kg, i.e. if the settler said he had 40 "heads", he was considered to have 40 AU.

			Coconut	
Plot area (hectare)	30	50	80	70
Livestock/beneficiary	45	40	31	36
Animal unit/hectare	1,5	0,8	0,38	0,51

Source: Research data, 2015.

In view of this, the Santa Emilia and Volta Grande settlements have an AU rate higher than the average for the micro-region, which is 0.60, showing that, even with a higher than average use capacity, the management practiced by the settlers has made it possible to recover from pasture degradation in both settlements. In the case of the Santa Emilia project, the animal unit in the settlement is twice the average practiced in the region and, even so, recovery of plant biomass was observed in the areas with the greatest intensity of degradation.

The Ilha do Coco settlement has a lower number of animals per hectare than the average for the micro-region, which is 0.52, and as a result intensive use of pasture and overgrazing are avoided. This becomes clear when analyzing that, over the years, there has been an intense recovery of all levels of degradation and an increase in plant biomass. Therefore, the management used, combined with the lower number of animals per hectare, has favored plant recovery in this settlement, consequently improving the vegetation index. In the Martins I settlement, the average number of animals per hectare is 0.51, very close to the regional average of 0.52. Although management has not significantly improved degradation in the most intense areas, it has made it possible to expand the areas without degradation, which indicates practices that are less aggressive to the environment.

The notion that the cultural practices and management employed by the beneficiaries are less aggressive and promote an improvement in pasture degradation is reinforced by the fact that 54% of those interviewed have never accessed any financial resources. Given this picture, it is clear that most of the settlers do not have the resources to invest in the plot, adopting practices such as soil fertilizer, for example, which would help in environmental recovery. In this way, proper management is one of the few alternatives for improving the vegetation condition of the pasture.

Credit is a very important tool for the families, as it makes it possible to improve the infrastructure of the plots, as well as investing in the purchase of animals and agricultural machinery. Only 46% of the families in the four settlements had access to some kind of financing, of which 86% invested the funds in restoring pastures and fences, buying cattle, restoring their houses and opening artesian wells. This situation shows that access to financial resources can help to improve and structure the plots.

Another important fact, when analyzing pasture degradation in these settlements, is that the improvement in vegetation is the result of the beneficiaries' lack of capital, who due to financial limitations are unable to exploit their plots more intensively. In this way, the recovery of the pasture is more a consequence of the lack of productive exploitation of the areas in the settlement than a direct and thought-out action on the part of the settlers in search of better conditions for the environment of which they are a part. In reality, what is happening is an increase in plant biomass, due to less intensive practices, which is not, in fact, a strategy or cultural practice for recovering degraded pastures.

In short, what was identified was the recovery of plant biomass and a reduction in pasture degradation in all the settlement projects surveyed. This was due to the fact that the cultural practices applied by the beneficiaries had a less intense impact on the vegetation. In addition, the reduction in the number of animals per hectare in the Ilha do Coco PA helped to recover the vegetation, revealing that reducing overgrazing also helps to restore degraded areas. What can be seen in these settlements is that, due to the economic restrictions that prevent more intensive use, there is a more rational use of natural resources, because even with a higher average number of animals, pastures have recovered in all the projects analyzed.

FINAL CONSIDERATIONS

The contribution that this study makes to the theme of agrarian reform is to provide a better understanding of the factors that condition permanence in rural agrarian reform settlements, given that a large number of studies have been identified on the reasons for evasion in settlements, but they do not address the issues addressed here. Thus, the data presented makes it possible to reflect on the socio-economic and environmental aspects of the policy of creating rural settlements.

As we can see, the issues raised are immersed in the reality experienced by rural land reform settlements in the state of Mato Grosso, and therefore the considerations presented, although similar to the reality of the vast majority of settlements in the country, need some caveats before being extrapolated to other realities.

It can be said that, when it comes to staying in rural settlements, there is a selection of beneficiaries, which means that the conditions of the environment and the lack of intervention by the public authorities promote a process of expulsion of families, with only the most adapted persisting - an adaptation almost always defined by access to external financial resources. This is because the rural settlements are in very poor infrastructural conditions, with little support from the government, no technical assistance and no access to credit. As a result, the beneficiaries of this public policy find themselves in economically and socially fragile conditions which, to a large extent, lead to evasion. On the other hand, settled farmers with some source of income outside the plots, or with some source of fixed monthly income, manage to stay in the settlements, unlike those who don't have these alternative incomes to agricultural production in the settlement. It is therefore a selection that the projects promote by making it possible for only those who have economic alternatives outside the plots to stay.

This work recognizes the settlement policy as a promoter of development because it guarantees workers access to the factor of production Land, expanding the freedoms of its beneficiaries. Thus, access to land is already an important element of this policy, as it creates the possibility for economically excluded groups, with low levels of education and no alternative employment in the urban-industrial sector, to remain in rural areas with security and dignity. However, the settlement policy has not managed to overcome the limits of the lack of access to another factor of production, Capital, here characterized by lines of financing, which in many cases makes it impossible for families to remain in the projects. Thus, guaranteed access to land is a significant and transformative achievement in the lives of rural workers, but it will not be able to fulfill its objectives of rural development and the emancipation of the beneficiaries if it is not accompanied by other elements such as lines of credit, technical assistance and adequate infrastructure.

Throughout this dissertation, it has been observed that pluriactivity is an important mechanism for guaranteeing workers' permanence in the settlement. It is associated with the ability to diversify sources of income and not depend exclusively on the plot to survive. This has become more evident over the years in the settlements, when there has been an increase in the number of beneficiaries who have bought their plot and who have a greater diversity of occupations, compared to the beneficiaries who were selected by INCRA and who only work in agricultural production. Pluriactivity provides families with financial stability, helping them to remain in rural areas. Because of this, understanding that rural settlements are not limited to access to land is an important factor in understanding how the beneficiaries of this policy organize themselves. It has become clear that the projects are not just places for agricultural production, but an environment in which a variety of activities are present, helping to maintain life in rural areas. Understanding that access to the plot is not limited solely to farming is an important step in the advancement of public policy.

Another important factor is the recovery of the pastures as a result of less intensive management. By analyzing satellite images, the environmental conditions in the initial and current periods of each settlement were compared and it was found that, in general, pasture degradation decreased in all the projects. The areas without degradation and the areas with less degraded classes increased, showing a more environmentally sustainable agricultural practice. This shows that the beneficiaries act as agents of environmental recovery, albeit indirectly and unintentionally. The practices they adopt exploit the vegetation less than those applied by conventional livestock farmers, which is evident in the improvement in the plant biomass index of the pastures.

In addition, attention is drawn to the use of livestock farming as a complement to the settler's income or savings, rather than as their main activity. In reality, the activities carried out in the settlements are complementary in terms of income generation, providing the beneficiaries with greater purchasing power with the resources obtained through other activities. In this sense, the settlers' living conditions are better than those of urban workers of the same economic level. In other words, the activities carried out by the beneficiaries, both inside and outside the plot, whether they generate income or are just for subsistence, create conditions that enable them to have access to a variety of foods, quality of life and an amount of resources that surpasses that of urban workers.

In all the projects studied, a precariousness was identified in relation to collective infrastructures, which are generally provided by the public authorities, whether at federal, state or municipal level, which also influences the question of permanence, insofar as their absence or precariousness makes these beneficiaries resort to other possibilities offered, often in urban centers. The precariousness of infrastructure influences the settlers' perception of the difficulties that permeate life in the settlement: in the Martins I project, which had a public school and a health center, for example, the main difficulties were not associated with distance from urban centers; however, in the Santa Emilia project, due to the lack of these collective structures, distance from the centers was one of the main difficulties experienced by the families interviewed. However, these issues cannot be considered determinants of evasion, since settlements with better structures can have higher evasion rates than those with worse conditions.

Finally, this work sought to present the characteristics that help beneficiaries of agrarian reform programs to remain in the country, an issue that is observed in the pluriactivity of the vast majority of these workers. The search for financial alternatives is not limited to the rural sphere, much less to productive activities on the plot, so this point is a determining factor in the decision to stay or leave. It is certain, therefore, that the guarantee of continuous resources is an important step in the advancement of this public policy, as is the understanding of rural settlements as places for not only agricultural development, but for activities that can guarantee the permanence of their residents through the development of multiple activities, including non-agricultural ones.

REFERENCES

ABRAMOVAY, R. Um novo contrato para a política de assentamentos. Oliva, P. M. (org.). **Economia Brasileira:** Perspectivas do Desenvolvimento. Sao Paulo: CAVC, 2005.

ALEIXO, D. N. S. **Changes in beneficiaries and ways of reoccupying plots in the Capelinha settlement, Conceiçâo de Macabu, RJ.** 2007 Dissertation Federal Rural University of Rio de Janeiro, Rio de Janeiro. 2007.

ALMEIDA, G. M.; LIMA, S. S. Geology of the Serra da Miaba through Landsat 8 images. [n.d.].

ALVARENGA, A. S.; MORAES, M. F. **Utilization of Landsat-8 images to characterize vegetation cover**, 2014.   Available   at:

<http://mundogeo.com/blog/2014/06/10/processamento-digital-de-imagens-landsat-8-para-obtencao-dos-indices-de-vegetacao-ndvi-e-savi-visando-a-caracterizacao-da-cobertura- vegetal-no-municipio-de-nova-lima-mg/>. Accessed on: 30 Apr. 2015

ALVES JR, G. T. O planejamento governamental e seus reflexos na estrutura fundiâria de Mato Grosso. **Caminhos de Geografia**, v. 4, n. 9, 2006.

ALVES, J.; FIGUEIREDO, A. M. R.; BONJOUR, S. C. M. Rural Settlements in Mato Grosso: An Analysis of Data from the Agrarian Reform Census. **Redalyc - Revistas Científicas de América Latina. el Caribe, Espana y Portugal,** v. 27, n. 39, p. 152-167, dec. 2009.

ANDRADE, R. et al. Monitoring pasture degradation processes using Spot Vegetation data. **Campinas: Embrapa Satellite Monitoring**, 2011.

ANDRADE, R. G. et al. Use of remote sensing techniques in the detection of pasture degradation processes. **Engenharia na Agricultura**, v. 21, n. 3, p. 234-243, 2013.

ARENAS-TOLEDO, J. M.; EPIPHANIO, J. C. N. Behavior of vegetation indices from three orbital sensors: a case study in the municipality of Sao Borja (RS). **XIII Brazilian Symposium on Remote Sensing**, v. 13, p. 741-748, 2007.

BARRETO, N. F. et al. Characterization and analysis of the degraded areas of the Santo Amaro Settlement. **Vértices**, v. 16, n. 3, p. 97-104, May 7, 2015.

BENTO, L. C. M. Reforma Agrària: a luta pela terra e a realidade do Assentamento Rural Rio das Pedras, Uberlândia/MG. **Enciclopédia Biosfera**, v. 5, n. 8, p. 18, 2009.

BERNARDES, J. A. **Shifting frontiers in the agrarian spaces of Mato Grosso's Araguaia Valley.** Available at: < http://www.nuclamb.geografia.ufrj.br/publicacoes/arquivos/arquivo 7.pdf>. Accessed on: May 20, 2014.

. **Modernization**: the logic of capital and the rights of the excluded. In: BERNARDES, J. A.; ARUZZO, R.C. (eds.). **New frontiers of technique in the Araguaia Valley**. Rio de Janeiro: Arquimedes Ediçoes, 2009.

BORGES, E. F. **Performance analysis of the NDVI and SAVI vegetation indices based on Aster images**. [s.d.].

BRAZIL. **Law No. 8629, of February 25, 1993**. Available at: <http://www.planalto.gov.br/ccivil_03/leis/L8629.htm>. Accessed in 2015.

BRUNO, R. The Land Statute: between conciliation and confrontation. **Studies in Society and Agriculture**, 2013.

CAMPOS, S. et al. Remote Sensing and Geoprocessing Applied to Land Use in Hydrographic Microbasins, Botucatu-SP. **Engenharia Agricola**, v. 24, n. 2, p. 431-435, 2004.

CARVALHO, H. M. DE. Compensatory rural settlement policy as a negation of agrarian reform. **Revista Nera**, n. 5, p. 113-122, 2012.

CASTRO, S. P. et. al. **A Colonizaçâo Oficial em Mato Grosso: "a nata e a borra da sociedade"**. Cuiabà: EDUFMT, 1994. 290 p.

COPPEL, I. A. F.; LLORENTE, E. H. **El Satélite Landsat: anàlisis visual de imâgenes obtenidas del sensor ETM+ Satélite Landsat**. [s.l.] Universidad de Valladolid, 2001.

CORRÊA, A. M. C.; FIGUEIREDO, N. M. S. Wealth, inequality and poverty: a profile of the Center-West region at the beginning of the 21st century. **Pesquisa e Debate**, v. 17, n. 1, 2006.

COSTA, L. F. C.; SANTOS, R. (Org.). **Politica e Reforma Agrària**. Rio de Janeiro: Mauad, 1998. 242 p.

DA COSTA, M. C. et al. **Avaliaçâo da dinàmica do uso da terra em uma região de fronteira agropecuària no estado de Mato Grosso**. [n.d.].

DA CUNHA, J. M. P. Migratory dynamics and the process of occupation of the Brazilian Midwest: the case of Mato Grosso. **Revista Brasileira de Estudos de Populaçao**, v. 23, n. 1, p. 87-107, 2013.

DA SILVA, T. P. Rural settlements in Càceres/MT: a space of life and peasant struggle. **Revista**

Eletrônica da Associaçao dos Geógrafos Brasileiros-Seçao Três Lagoas-MS, v. 1, n. 15, 2012.

DA VEIGA, J. B. et al. Socioeconomic diagnosis of the residents of the Aruma settlement, Apiacàs, Mato Grosso. **Revista da Universidade Vale do Rio Verde**, v. 12, n. 2, p. 423-433, 2014.

DAVID, M. B. DE A.; WANIEZ, P.; BRUSTLEIN, V. Atlas dos beneficiârios da reforma agrària. **Estudos Avançados**, v. 11, n. 31, p. 51-68, 1997.

; ; . **Social and demographic situation of land reform beneficiaries**: an atlas. 1998.

DE CARVALHO DORES, E. F. G.; DE-LAMONICA-FREIRE, E. M. Contamination of the aquatic environment by pesticides. Case study: water used for human consumption in Primavera do Leste, Mato Grosso - preliminary analysis. **Chem. Nova,** v. 24, n. 1, p. 27-36, 2001.

DE OLIVEIRA, I. L. et al. **Family farming and social reproduction strategies in rural settlements in Mato Grosso: the case of the Fazenda Esperança settlement in Rondonópolis-MT.** In: XX Encontro Nacional de Geografia Agrària, 2012. Available at: <http://www.lagea.ig.ufu.br/xx1enga/anais_enga_2012/eixos/1166_1.pdf>. Accessed on: May 14, 2015

DELGADO, G. C. The agrarian question in Brazil, 1950-2003. **Social issues and social policies in contemporary Brazil.** Brasilia: IPEA, p. p51-90, 2005.

. Expansion and modernization of the agricultural sector in the post-war period: a study of agrarian reflection. **Estudos avançados**, v. 15, n. 43, p. 157-172, 2001.

DO NASCIMENTO, A. F. et al. Land Cover Classification Using the Free Programs: InterIMAGE, WEKA and QuantumGIS. In: XVI Brazilian Symposium on Remote Sensing. **Proceedings**... Foz do Iguaçu, 2013 Available at: <http://www.dsr.inpe.br/sbsr2013/files/p1453.pdf>. Accessed on: May 14, 2015

DOMINGUES, J. M. The dialectics of conservative modernization and the new history of Brazil. **Dados**, v. 45, n. 3, p. 459-482, 2002.

ESQUERDO, V. F. D. S.; BERGAMASO, S. Agrarian reform and rural settlements: perspectives and challenges. **NEAD**, p. 23, [n.d.]. Available at:

<http://portal.mda.gov.br/portal/nead/arquivos/download/Artigo   012.pdf?file   id=9145714> Accessed 2015.

FERNANDES, B. M. DATALUTA-Banco de Dados da Luta pela Terra. **Revista Nera**, n. 3, p. 7-27, 2012.

FERNANDEZ, A. J. C. **From the Cerrado to the Amazon: the social structures of the soy**

economy in Mato Grosso. 2007. 262 p. Dissertation (Master's Degree) - Federal University of Rio Grande do Sul, Porto Alegre, 2007.

______; FERREIRA, E. C. The socio-economic   impacts of rural settlements in Mato Grosso. In: MEDEIROS, L. S.; LEITE, S. (Org.). **Rural Settlements: Social Change and Regional Dynamics.** Rio de Janeiro: Mauad, 2004. p. 187-228.

FERRANTE, V. L. S. B.; BARONE, L. A.; DUVAL, H. C. Experiências de reforma agrària: bloqueios e perspectivas de desenvolvimento rural. **Lutas e Resistências**, v. I, n. 1, p. 17, jul. 2006.

FERREIRA, E. C.; FERNANDEZ, A. J. C.; SILVA, E. P. The reconstruction of rural settlements in Mato Grosso. In: MEDEIROS, L. S.; LEITE, S. (Org.). **A Formaçao dos Assentamentos Rurais no Brasil**. Porto Alegre/Rio de Janeiro: Ed. Universidade/UFRGS/CPDA, 1999. p. 197-231.

FRANÇA, C. G.; SPAROVEK, G. (Coords.). **Settlements in debate.** Brasilia: NEAD, 2005. 300 p.

GAO, Q. et al. Grassland degradation in Northern Tibet based on remote sensing data. **Journal of geographical sciences**, v. 16, n. 2, p. 165-173, 2006.

GOMES, D. et al. **Comparative evaluation of atmospheric correction of Landsat images using MODTRAN and Dark Object Subtraction**. [n.d.].

GOMES, M.; SANTOS, M. **Socio-economic ecological zoning**: socio-economic ecological diagnosis of the state of Mato Grosso and technical assistance in the formulation of the 2nd approach. Cuiabà: Mato Grosso State Government/Seplan/Bird, 2000.

GONÇALVES, R. Assentamentos como pactos de (des) interesses nos governos democrâticos. **Lutas Sociais-Desde 1996-ISSN 1415-854X**, n. 15/16, p. 184-198, 2006.

GOUVEIA, R. G. L. DE et al. Diagnosis on the administration settlements' families in Tangara da Serra-MT: the case of the project credit of Vale do Sol II. **Current Agricultural Science and Technology**, v. 18, n. 4, 2013.

GUANZIROLI, C. ; BRUNO, R.; MEDEIROS, L. (Coords.). **Percentages and Causes of Evasion in Rural Settlements.** Ministry of Agrarian Development: National Institute for Colonization and Agrarian Reform. Brasilia, 2001.

;;. Percentages and causes of evasion in rural settlements. **Cadernos da Terra N series**, v. 9, 2001.

HEREDIA, B. et al. Analysis of the regional impacts of agrarian reform in Brazil. **Studies in Society and Agriculture**, v. 1, 2013.

PALMEIRA, M.; LEITE, S. P. Sociedade e economia do "agronegócio" no Brasil. **Revista**

Brasileira de Ciências Sociais, v. 25, n. 74, p. 159-176, 2010.

HOFFMAN, R.; NEY, M. G. **Estrutura fundiària e propriedade agricola no Brasil, grandes regiôes e unidades da federaçâo**. [s.l.] Brasilia: Ministry of Agrarian Development, 2010.

IMEA - Institute of Applied Economics of Mato Grosso. Available at: < http://imea.com.br/upload/caracterizacaoBovinocultura.pdf>. Accessed in 2015.

INCRA - NATIONAL INSTITUTE FOR COLONIZATION AND AGRARIAN REFORM. Available at: <www.incra.org.br>. Accessed in 2015.

. Journal December 2010, n° 02, 2010. Available at: <

http://www.incra.gov.br/media/servicos/publicacao/livros_revistas_e_cartilhas/jornal_incra_2_7_01_2011.pdf>. Accessed on: June 1, 2014.

. **Execution Standard No. 45.** Provides for procedures for selecting candidates for the National Agrarian Reform Program. 2005

SILVA, J. ; HAMULAK, T. M. ; RIBEIRO, S. R. A. . Comparison between images with and without atmospheric correction using the Image Quality Index. **Anais XVI Simpòsio Brasileiro de Sensoriamento Remoto**, p. 4563-4569, 2013.

KAGEYAMA, A.; BERGAMASO, S.; OLIVEIRA, J. **Os assentamentos rurais no Censo Agropecuârio de 2006** Sao Paulo: UNICAMP, 2006.

KALAF, R.; CARDOSO, P. V.; CRUZ, C. B. M. **Landsat 8**: Advances for mesoscale mapping. [n.d.].

KICHEL, A. N.; MIRANDA, C. H. B.; ZIMMER, A. H. Pasture degradation and beef cattle production with agriculture x livestock integration. **Simpòsio de Produçâo de Gado de Corte**, v. 1, p. 201-234, 1999.

LAMERA, J. A.; FIGUEIREDO, A. M. R. Rural Settlements in Mato Grosso. 46th Congress, July 20-23, 2008, Rio Branco, Acre, Brazil. **Proceedings**. Brazilian Society of Economics, Administration and Rural Sociology (SOBER), 2008. Available at: <http://ideas.repec.org/p/ags/sbrfsr/108160.html>. Accessed on: May 14, 2015

ZAVALA, A. Z. Data envelopment analysis in the study of efficiency in rural settlements in the State of Mato Grosso. 46th Congress, July 20-23, 2008, Rio Branco, Acre, Brazil. **Proceedings**. Brazilian Society of Economics, Administration and Rural Sociology (SOBER), 2008. Available at: <https://ideas.repec.org/p/ags/sbrfsr/108137.html>. Accessed on: May 14, 2015

LEITE, S. et. al. (Coord.) Regional impacts of agrarian reform in Brazil: political, economic and social aspects. **Agrarian reform and sustainable development. Brasilia: Paralelo**, v. 21, 2000.

LEVIN, Jack. **Estatistica Aplicada a Ciências Humanas**. 2nd Ed. Sao Paulo: Editora Harbra Ltda, 1987.

LIMA, G. C. et al. Evaluation of vegetation cover by the normalized difference vegetation index (NDVI). **Revista Ambiente & Agua-An Interdisciplinary Journal of Applied Science: v**, v. 8, n. 2, 2013.

MACEDO, M. C. M.; KICHEL, A. N.; ZIMMER, A. H. **Degradação e alternativas de recuperaçao e renovaçao de pastagens**. [Embrapa Gado de Corte, 2000.

MACHADO, L. E. G.; CEDRO, D. B. Evoluçao do Uso Agropecuârio no Periodo de 1975 a 2008 no município de Barra do Garças-MT. **XIII SBGFA**, 2009.

MAIA, G. S.; KHAN, A. S.; SOUZA, E. P.. Assessing the impact of the federal agrarian reform program in Ceará: a case study. **ECONOMIA APLICADA**, v. 17, n. 3, p. 379398, 2013.

MARAFON, G. J. Industrializaçao da agricultura e formaçao do complexo agroindustrial no Brasil. **GEO UERJ-Revista do Departamento de Geografia da UERJ**, n. 3, 1998.

MARQUES, V. P.; DEL GROSSI, M. E.; FRANÇA, C. G. The 2006 Census and agrarian reform: methodological aspects and first results. 2012.

MARTINE, G. The trajectory of agricultural modernization: who benefits? **Lua Nova: Revista de Cultura e Politica**, n. 23, p. 7-37, 1991.

MARTINS, J. S. **O cativeiro da terra**. 9. ed. Sao Paulo: Contexto, 2013. 283 p.

MATTEI, L. The Brazilian agrarian reform: evolution of the number of settled families in the post-redemocratization period of the country. **Estudos Sociedade e Agricultura**, v. 2, 2013.

MEDEIROS, L. et. al. (Org.). **Rural settlements: a multidisciplinary view.** Sao Paulo: Editora da Universidade Estadual Paulista, 1994. 325 p.

; LEITE, S. The regional impacts of rural settlements: economic, political and social dimensions. **CPDA/Debates**, v. 4, 1997.

Rural Settlements: Social Change and Regional Dynamics. Rio de Janeiro: Mauad, 2004. 307 p.

MELO, E. T.; SALES, M. C. L.; DE OLIVEIRA, J. G. B. Aplicação do indice de Vegetaçao por Diferença Normalizada (NDVI) para análise da degração ambiental na Microbacia Hidrografica do Riacho dos Cavalos, Crateùs-CE. **Raega-O Espaço Geogràfico em Anàlise**, v. 23, [n.d.].

MOREIRA, E.; TARGINO, I.; MENEZES, M. Espaço agràrio, movimentos sociais e a acção fundiària na zona canavieira do Nordeste. **Cadernos de Estudos Sociais**, v. 19, n. 2, p. 201226,

2003.

MORENO, G. The historical process of access to land in Mato Grosso. **Geosul**, v. 14, n. 27, p. 67-90, 1999.

NASCIMENTO, C. R. **Atmospheric correction of AVHRR/NOAA sensor images using atmospheric products from the MODIS/TERRA sensor**. Dissertation (Master's Degree) - State University of Campinas, Campinas. 2006.

NAVARRO, Z. Dilemmas of a protagonist in the struggle for land. In: COSTA, L. F. C.; SANTOS, R. (Org.). **Politica e Reforma Agrària**. Rio de Janeiro: Mauad, 1998. p. 181-184.

OLIANI, L. O.; PAIVA, C.; ANTUNES, A. F. B. Utilizaçao de Softwares Livres de Geoprocessamento para Gestao Urbana em Municipios de Pequeno e Mèdio Porte. **IV Brazilian Symposium on Geodetic Sciences and Geoinformation Technologies**, p. 01-08, 2012.

ORLANDI, M.; DE LIMA, J. F. The effective occupation of the territory and the growth of economic activities in Mato Grosso-1980 to 2007. **Anais: Encontros Nacionais da ANPUR**, v. 14, 2013.

QUADRO, F. Sofwtares livres no Ensino. **Revista FOSSGIS Brasil,** p. 9-12, 2011.

RAMOS, R. et al. Application of the normalized difference vegetation index (NDVI) in the evaluation of degraded areas and potential for conservation units.**III Simpòsio Brasileiro de Ciências Geodésicas e Tecnologias da Geoinformaçâo**, p. 001-006, 2010.

RIBEIRO, M. M. C. **Models of agrarian reform: evasion and permanence in rural settlements in the state of Tocantins**. 2009. 104 p. Dissertation (Master's Degree) - Federal University of Viçosa, Viçosa, 2009.

ROCHA, H. F. Territorial dispute, conceptualization and actuality of Agrarian Reform in Brazil. **GeoGraphos: Digital Journal for Students of Geography and Social Sciences**, v. 4, n. 50, p. 433-462, 2013.

RUDORFF, B.; MOREIRA, M.; ALVES, M. Remote sensing applied to agriculture. **National Institute for Space Research - INPE**, p. 19, 2002.

SANCHES, I. D. et al. Comparative analysis of three methods of atmospheric correction of Landsat 5-TM images to obtain surface reflectance and NDVI. In: Simpòsio Brasileiro de Sensoriamento Remoto, 15., 2011, Curitiba. **Proceedings...** Sao José dos Campos: INPE,2011, 2011Available at: <http://bibdigital.sid.inpe.br/rep-

/dpi.inpe.br/marte/2011/07.11.12.17>. Accessed on: May 14, 2015

SAUER, S. The meaning of agrarian reform settlements in Brazil. In: FRANÇA, C. G.; SPAROVEK, G. (Coord.). **Assentamentos em debate.** Brasilia: NEAD, 2005. p. 57-74.

SILVA, M. J. DA; SATO, M. T. Territories in tension: mapping socio-environmental conflicts in the state of Mato Grosso-Brazil. **Ambiente & Sociedade**, v. 15, n. 1, p. 122, 2012.

SOUSA, M. S. Avaliaçao das Imagens CBERS/CCD para o Mapeamento de Areais no sudoeste De Goiàs - DOI 10.5216/bgg. v27i2. 2660. **Boletim Goiano de Geografia**, v. 27, n. 2, p. 115-137, 2007.

SOUZA, I. C. **In search of the "pure movement". Organization, representation and local political dispute.** 2005. 282 p. Thesis (Doctorate) - Federal Rural University of Rio de Janeiro, Rio de Janeiro, 2005.

SPAROVEK, G. **A qualidade dos assentamentos da reforma agrària brasileira.** [s.l.] Pâginas & Letras, 2003.

TANAJURA, E. L. X.; ANTUNES, M. A. H.; UBERTI, M. S. Evaluation of Vegetation Indices for the Discrimination of Agricultural Targets in Satellite Images. **Brazilian Symposium on Remote Sensing**, v. 12, 2005.

TEIXEIRA, J. C. Modernization of agriculture in Brazil: economic, social and environmental impacts. **Revista Eletrônica da Associaçâo dos Geògrafos Brasileiros**, v. 2, n. 2, p. 2142, 2005.

THÉRY, H. et al. **Dataluta article.** [n.d.].

USGS - UNITED STATES GEOLOGICAL SURVEY. Available at: < http://www.usgs.gov/>. Accessed 2015.

UCHOA, H. N.; FERREIRA, P. R. Geoprocessing with free software. **Electronic publication.** Rio de Janeiro, 2004.

VIGANÓ, H. A.; FRANCA-ROCHA; BORGES, E. F. Performance analysis of NDVI and SAVI vegetation indices from Aster images. **Brazilian Symposium on Remote Sensing**, p. 1828-1834, 2011.

WEISSHEIMER, C. et al. Participatory Rapid Diagnosis of production systems in an Agrarian Reform Settlement in the Cerrado region of the Upper Xingu Basin. **Revista Brasileira de Agroecologia**, v. 2, n. 2, 2007.

I want morebooks!

Buy your books fast and straightforward online - at one of world's fastest growing online book stores! Environmentally sound due to Print-on-Demand technologies.

Buy your books online at
www.morebooks.shop

Kaufen Sie Ihre Bücher schnell und unkompliziert online – auf einer der am schnellsten wachsenden Buchhandelsplattformen weltweit! Dank Print-On-Demand umwelt- und ressourcenschonend produzi ert.

Bücher schneller online kaufen
www.morebooks.shop

info@omniscriptum.com
www.omniscriptum.com

Printed by Books on Demand GmbH, Norderstedt / Germany